Fractional Calculus with its Applications in Engineering and Technology

Synthesis Lectures on Mechanical Engineering

Synthesis Lectures on Mechanical Engineering series publishes 60–150 page publications pertaining to this diverse discipline of mechanical engineering. The series presents Lectures written for an audience of researchers, industry engineers, undergraduate and graduate students.

Additional Synthesis series will be developed covering key areas within mechanical engineering.

Fractional Calculus with its Applications in Engineering and Technology
Yi Yang and Haiyan Henry Zhang
2019

Essential Engineering Thermodynamics: A Student's Guide
Yumin Zhang
2018

Engineering Dynamics
Cho W.S. To
2018

Solving Practical Engineering Problems in Engineering Mechanics: Dynamics
Sayavur I. Bakhtiyarov
2018

Solving Practical Engineering Mechanics Problems: Kinematics
Sayavur I. Bakhtiyarov
2018

C Programming and Numerical Analysis: An Introduction
Seiichi Nomura
2018

Mathematical Magnetohydrodynamics
Nikolas Xiros
2018

Design Engineering Journey
Ramana M. Pidaparti
2018

Introduction to Kinematics and Dynamics of Machinery
Cho W. S. To
2017

Microcontroller Education: Do it Yourself, Reinvent the Wheel, Code to Learn
Dimosthenis E. Bolanakis
2017

Solving Practical Engineering Mechanics Problems: Statics
Sayavur I. Bakhtiyarov
2017

Unmanned Aircraft Design: A Review of Fundamentals
Mohammad Sadraey
2017

Introduction to Refrigeration and Air Conditioning Systems: Theory and Applications
Allan Kirkpatrick
2017

Resistance Spot Welding: Fundamentals and Applications for the Automotive Industry
Menachem Kimchi and David H. Phillips
2017

MEMS Barometers Toward Vertical Position Detecton: Background Theory, System Prototyping, and Measurement Analysis
Dimosthenis E. Bolanakis
2017

Engineering Finite Element Analysis
Ramana M. Pidaparti
2017

Fractional Calculus with its Applications in Engineering and Technology
Yi Yang and Haiyan Henry Zhang

ISBN: 978-3-031-79624-1   paperback
ISBN: 978-3-031-79625-8   ebook
ISBN: 978-3-031-79626-5   hardcover

DOI 10.1007/978-3-031-79625-8

A Publication in the Springer series
SYNTHESIS LECTURES ON MECHANICAL ENGINEERING

Lecture #17
Series ISSN
Print 2573-3168   Electronic 2573-3176

Fractional Calculus with its Applications in Engineering and Technology

Yi Yang and Haiyan Henry Zhang
Purdue University

SYNTHESIS LECTURES ON MECHANICAL ENGINEERING #17

ABSTRACT

This book aims to provide the basic theory of fractional calculus and its applications based on practical schemes and approaches, illustrated with applicable engineering and technical examples, especially focusing on the fractional-order controller design. In the development of this book, the essential theorems and facts in the first two chapters are proven with rigorous mathematical analyses. In addition, the commonly used definitions of Grünwald-Letnikov, Riemann-Liouville, Caputo, and Miller-Ross fractional derivatives are introduced with their properties proved and linked to fractional-order controller design. The last chapter presents several enlightening scenarios of fractional-order control designs, for example, the suppression of machining chatter, the nonlinear motion control of a multilink robot, the simultaneous tracking and stabilization control of a rotary inverted pendulum, and the idle speed control of an internal combustion engine (ICE).

KEYWORDS

fractional calculus, fractional-order controller design, Grünwald-Letnikov, Riemann-Liouville, Caputo, Miller-Ross fractional derivatives, machining chatter, multilink robot, rotary inverted pendulum, idle speed control

Contents

Preface

The principal elements of fractional calculus are instituted with mathematical proofs and practical illustrations in the engineering and technology of this book. The book is particularly valuable for college students and control engineers who are interested in fractional calculus theory and its applications, especially in control system engineering.

This book is divided into three chapters. Chapter 1 introduces the fundamental definitions and theorems in fractional calculus. Chapter 2 provides the time-domain and frequency-domain analysis for the fractional-order control system, including its stability, controllability, and observability analysis. Subsequently, a series of constraints for designing an optimal fractional-order controller are discussed. Finally, several practical control problems with continuous or discrete fractional-order controllers are studied in Chapter 3 as a demonstration of the effectiveness of fractional-order controllers.

Yi Yang and Haiyan Henry Zhang
March 2019

Acknowledgments

Fractional calculus and its applications in engineering and technology is a novel topic. This book would not be possible without the forerunners' contributions from academia and industry. We would like to thank Prof. Mark French at Purdue University for his helpful discussions on fractional calculus. We would also like to thank Dr. Yuequan Wan for his collaborative work on the design and optimization of the fractional-order controllers, extended from the integer-order control laws derived by Dr. Haiyan Henry Zhang in machining chatter and multilinked robot controls. Last but not least, we are also grateful to Morgan & Claypool Publishers for their foresight in emerging engineering and technology applications of fractional calculus and generous support of this book's publication.

Yi Yang and Haiyan Henry Zhang
March 2019

Introduction

In recent years, fractional calculus is finding its applications in more and more fields of engineering and technology. It is an effective means for superior modeling and robust control of interdisciplinary systems, distributive systems, and complex systems, as well as for the characteristic identification of flexible circuits, transient analysis of ordinary engineering systems, study of rheological properties, etc. This book introduces the basics of fractional calculus in terms of functions employed to construct the fractional derivatives and integrals, analytical solutions to linear or linearized fractional differential equations, and numerical solutions to general form fractional differential equations. Its control design is also derived in frequency domain, pole placement, and state space, as illustrated in several real-world examples of fractional calculus.

With its emerging applications in engineering and technology, fractional calculus has caught more and more attention in modeling and control. The basics and tools for fractional calculus are feasible and practical as a part of the engineering and technology trainings for undergraduate students. This short book introduces the systematic theory of fractional calculus with definitions and their explanations, theorems and their proofs, as well as the analytical solutions and numerical simulations of engineering and technology problems.

CHAPTER 1

Preliminary Tools of Fractional Calculus

1.1 SPECIAL FUNCTIONS

This section introduces fundamental functions which are developed in conventional calculus and play important roles in the establishment of fractional calculus. One of them is gamma function, which is the extension of the integer's factorial. Beta function is a well-known function which bridges the convolution operator and the fractional calculus theory. Using the definitions of gamma function and beta function, many complicated calculations and formulas can be simplified. The Mittag–Leffler function is a generalization of the exponential function, which can be employed as a tool for analytically solving a specific type of fractional ordinary differential equations. Other special functions are also briefly introduced at the end of this section.

1.1.1 GAMMA FUNCTION

The definition of the gamma function is given by the integral expressed as

$$\Gamma(z) = \int_0^\infty e^{-t} t^{z-1} dt, \tag{1.1}$$

where in Equation (1.1) the domain of convergence is the whole complex plane except for the non-positive integers. A brief proof illustrating the convergence of integral in Equation (1.1) in the open right half-complex plane is given as follows. And its convergence in the open left-half plane excluding the non-positive integer poles will be self-evident when a useful property of Equation (1.1) is introduced afterward.

Proof. Suppose $z = x + iy$, where $x, y \in \mathbb{R}$. We have

$$\Gamma(z) = \int_0^\infty e^{-t} t^{x+iy-1} dt = \int_0^\infty e^{-t} t^{x-1} e^{iy \log t} dt; \tag{1.2}$$

since $\left|e^{iy \log t}\right| = 1$, it implies $|\Gamma(z)| \leq \int_0^\infty e^{-t} t^{x-1} dt$. By Cauchy's comparison theorem, the convergence of $\int_0^\infty e^{-t} t^{x-1} dt$ with $x > 0$ leads to the convergence of the gamma function when

$x > 0$. Indeed,

$$\int_0^\infty e^{-t}t^{x-1}dt = \int_0^1 e^{-t}t^{x-1}dt + \int_1^\infty e^{-t}t^{x-1}dt$$
$$< \int_0^1 t^{x-1}dt + \int_1^\infty e^{-t}t^{x-1}dt. \tag{1.3}$$

It is obvious that $\int_0^1 t^{x-1}dt$ is convergent when $x > 0$, and $\int_1^\infty e^{-t}t^{x-1}dt$ is convergent for any $x \in \mathbb{R}$. Thus, it concludes that gamma function is convergent at least in the open right half-complex plane.   □

Gamma function holds several useful properties.

- Property 1:

$$\Gamma(z + 1) = z\Gamma(z). \tag{1.4}$$

- Property 2:

$$\Gamma(1 - z)\Gamma(z) = \frac{\pi}{\sin \pi z}. \tag{1.5}$$

- Property 3 (Legendre Duplication Formula):

$$\Gamma(z)\Gamma\left(z + \frac{1}{2}\right) = 2^{1-2z}\sqrt{\pi}\Gamma(2z). \tag{1.6}$$

Proof of Property 1. It is verified by integrating by parts

$$\Gamma(z + 1) = \int_0^\infty e^{-t}t^z dt = -e^{-t}t^z\Big|_{t=0}^{t=\infty} + z\int_0^\infty e^{-t}t^{z-1}dt = z\Gamma(z).$$

□

By substituting the Taylor's series of the exponential term in the gamma integral into the definition, the gamma integral can be expressed as the sum of an entire function and an infinite series shown as follows [Podlubny, 1999, Uchaikin, 2013]:

$$\Gamma(z) = \sum_{k=0}^\infty \frac{(-1)^k}{k!}\frac{1}{k+z} + \int_1^\infty e^{-t}t^{z-1}dt.$$

Property 1 implies the possibility of conducting analytic continuation of gamma function from the right half-open complex plane to the open left-half complex domain, excluding the

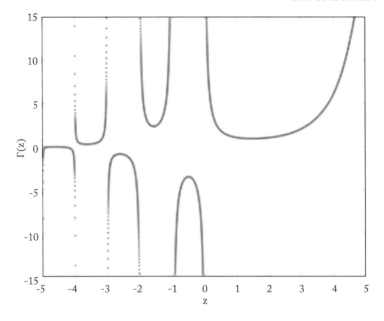

Figure 1.1: Sketch of the gamma function.

points with non-positive integer poles by using $\Gamma(z) = \frac{\Gamma(z+1)}{z}$. Therefore, the convergence domain of gamma function is the entire complex plane except non-positive integer poles. Gamma function can be visualized in Figure 1.1 if setting $z \in \mathbb{R}$.

Proof of Property 2 and Property 3 will be provided when the beta function is introduced in the next section.

1.1.2   BETA FUNCTION

Beta function can be defined as

$$B(z, w) = \int_0^1 \tau^{z-1}(1-\tau)^{w-1}\, d\tau, \quad Re(z) > 0, \quad Re(w) > 0. \tag{1.7}$$

Beta function is a specific case of the convolution of two functions t^{z-1} and t^{w-1}, as shown in the following equation when the convolution integral takes value at $t = 1$, i.e.,

$$h_{z,w}(t) = \int_0^t \tau^{z-1}(t-\tau)^{w-1}\, d\tau. \tag{1.8}$$

Obviously, $B(z, w) = h_{z,w}(1)$, and the relationship between beta function and gamma function can be presented in Equation (1.9) [Monje et al., 2010, Podlubny, 1999, Xue, 2017]:

$$B(z, w) = \frac{\Gamma(z)\Gamma(w)}{\Gamma(z + w)}. \tag{1.9}$$

Proof of Equation (1.9). The Laplace transform of the convolution integral (Equation (1.8)) is given as the product of the two functions' convolution, i.e.,

$$
\begin{aligned}
\mathcal{L}\{h_{z,w}(t)\} &= \mathcal{L}\{L\}\{t^{z-1}\} \cdot \mathcal{L}\{L\}\{t^{w-1}\} \\
&= \int_0^\infty e^{-st} t^{z-1} dt \cdot \int_0^\infty e^{-st} t^{w-1} dt \\
&= \frac{1}{s^z} \int_0^\infty e^{-st}(st)^{z-1} d(st) \cdot \frac{1}{s^w} \int_0^\infty e^{-st}(st)^{w-1} d(st) \\
&= \frac{\Gamma(z)\Gamma(w)}{s^{z+w}}.
\end{aligned}
$$

The inverse Laplace transform gives

$$h_{z,w}(t) = \mathcal{L}^{-1}\left\{\frac{\Gamma(z)\Gamma(w)}{s^{z+w}}\right\} = \frac{\Gamma(z)\Gamma(w)}{\Gamma(z+w)} \cdot t^{z+w-1}.$$

Taking $t = 1$ leads to the following relation:

$$B(z, w) = h_{z,w}(1) = \frac{\Gamma(z)\Gamma(w)}{\Gamma(z+w)}.$$

$\square$

Next, we are able to present the proof for the gamma function's Property 2 and Property 3 using the relation in Equation (1.9).

Proof of Property 2 of $\Gamma(z)$. Let's first consider $0 < Re(z) < 1$:

$$\Gamma(z)\Gamma(1-z) = B(z, 1-z) = \int_0^1 \left(\frac{u}{1-u}\right)^{z-1} \frac{du}{1-u}.$$

If the transformation $v = \frac{u}{1-u}$ is applied to the above formula, we have

$$\Gamma(z)\Gamma(1-z) = \int_0^\infty \frac{v^{z-1}}{1+v} dv.$$

Using Cauchy–Goursat theorem in complex analysis, we are able to convert the above integration into a contour integration with contour plotted in Figure 1.2. Let

$$f(s) = \frac{s^{z-1}}{1+s}.$$

For the radius $R > 1$, the contour integration can be expressed as

$$\oint f(s)\, ds = 2\pi i \left[Res\, f(s)\right]_{s=e^{\pi i}} = -2\pi i e^{i\pi z}.$$

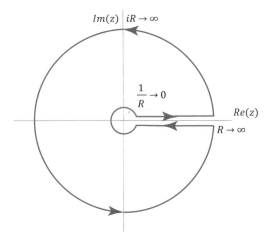

Figure 1.2: Integration contour.

Since $0 < Re\,(z) < 1$, the contour integrations along the outer circle and the inner circle are vanished when $R \to \infty$, which means that the whole contour integration is equal to the sum of the integration along the upper cutting edge and that along the lower cutting edge in the positive real axis, i.e.,

$$\int_0^\infty \frac{v^{z-1}}{1+v}\, dv + \int_\infty^0 \frac{\left(v e^{2\pi i}\right)^{z-1}}{1+v}\, dv = \oint f(s)\, ds = -2\pi i e^{i\pi z}$$

$$\int_0^\infty \frac{v^{z-1}}{1+v}\, dv = \frac{-2\pi i e^{i\pi z}}{1-e^{2\pi i z}} = \frac{\pi}{\sin \pi p}.$$

For the general case such that $m < Re\,(z) < m+1$ and m is any integer, we can set $z = \gamma + m$; Property 1 implies the following relation:

$$\Gamma\,(z)\,\Gamma\,(1-z) = (-1)^m\,\Gamma\,(\gamma)\,\Gamma\,(1-\gamma) = \frac{(-1)^m \pi}{\sin(\pi \gamma)} = \frac{\pi}{\sin(\pi z)}.$$

$\square$

Proof of Property 3 of $\Gamma(z)$. Since

$$B(z,z) = \int_0^1 [t(1-t)]^{z-1}\, dt = 2\int_0^{\frac{1}{2}} [t(1-t)]^{z-1}\, dt.$$

Substituting $s = 4t(1 - t)$ into the equation above yields:

$$B(z, z) = 2^{1-2z} \int_0^1 s^{z-1}(1 - s)^{-1/2} ds = 2^{1-2z} B\left(z, \frac{1}{2}\right).$$

Utilizing Equation (1.9) gives:

$$B(z, z) = \frac{\Gamma(z)\Gamma(z)}{\Gamma(2z)} = 2^{1-2z} B\left(z, \frac{1}{2}\right) = 2^{1-2z} \frac{\Gamma(z)\Gamma\left(\frac{1}{2}\right)}{\Gamma\left(z + \frac{1}{2}\right)}.$$

Given $\Gamma\left(\frac{1}{2}\right) = \sqrt{\pi}$ from Property 2, we have

$$\Gamma(z)\Gamma\left(z + \frac{1}{2}\right) = 2^{1-2z} \sqrt{\pi} \Gamma(2z).$$

$\square$

1.1.3   MITTAG–LEFFLER FUNCTION

Two parameter Mittag–Leffler function is as follows:

$$E_{\lambda,\mu}(z) = \sum_{k=0}^{\infty} \frac{z^k}{\Gamma(\lambda k + \mu)}, \quad (\lambda > 0, \quad \mu > 0). \qquad (1.10)$$

From the definition shown in Equation (1.10), the exponential function e^z is a special case of the Mittag–Leffler function when $\lambda = \mu = 1$.

Laplace transform of the Mittag–Leffler function is a key for finding solutions of the fractional-order differential equations. Laplace transforming Equation (1.10) yields [Podlubny, 1999]:

$$\int_0^{\infty} e^{-st} t^{\lambda k + \mu - 1} E_{\lambda,\mu}^{(k)}\left(at^{\lambda}\right) dt = \frac{k! s^{\lambda - \mu}}{\left(s^{\lambda} - a\right)^{k+1}}, \quad \left(Re(s) > |a|^{1/\lambda}\right). \qquad (1.11)$$

Proof of Equation (1.11).

$$\mathcal{L}\left\{t^{\lambda k+\mu-1}E_{\lambda,\mu}^{(k)}(at^\lambda)dt\right\} = \int_0^\infty e^{-st}t^{\lambda k+\mu-1}\sum_{m=k}^\infty \frac{\Gamma(m+1)}{\Gamma(m-k+1)}\cdot\frac{(at^\lambda)^{m-k}}{\Gamma(\lambda m+\mu)}dt$$

$$= \sum_{m=k}^\infty \frac{\Gamma(m+1)}{\Gamma(m-k+1)}\cdot\frac{a^{m-k}}{\Gamma(\lambda m+\mu)}\int_0^\infty e^{-st}t^{\lambda m+\mu-1}dt$$

$$= \sum_{m=k}^\infty \frac{\Gamma(m+1)}{\Gamma(m-k+1)}\cdot\frac{a^{m-k}}{\Gamma(\lambda m+\mu)}s^{1-(\lambda m+\mu)}$$

$$\cdot s^{-1}\int_0^\infty e^{-st}(st)^{\lambda m+\mu-1}d(st)$$

$$= \sum_{m=k}^\infty \frac{\Gamma(m+1)}{\Gamma(m-k+1)}\cdot\frac{a^{m-k}s^{-(\lambda m+\mu)}}{\Gamma(\lambda m+\mu)}\Gamma(\lambda m+\mu)$$

$$= \sum_{m=k}^\infty \frac{\Gamma(m+1)}{\Gamma(m-k+1)}a^{m-k}s^{-(\lambda m+\mu)}$$

$$= k!s^{-\lambda k-\mu}\sum_{i=0}^\infty \binom{k+i}{i}\left(as^{-\lambda}\right)^i$$

$$= k!s^{-\lambda k-\mu}\sum_{i=0}^\infty \frac{(k+i)(k+i-1)\cdots(k+1)}{i!}\left(as^{-\lambda}\right)^i$$

$$= k!s^{-\lambda k-\mu}\sum_{i=0}^\infty \frac{(-k-1)(-k-2)\cdots(-k-i)}{i!}(-1)^i\left(as^{-\lambda}\right)^i$$

$$= k!s^{-\lambda k-\mu}\sum_{i=0}^\infty \binom{-k-1}{i}\left(-as^{-\lambda}\right)^i$$

$$= \frac{k!s^{-\lambda k-\mu}}{(1-as^{-\lambda})^{k+1}} = \frac{k!s^{\lambda-\mu}}{(s^\lambda-a)^{k+1}}.$$

$\square$

By setting λ, μ or s to appropriate values in Equation (1.11), some corollary integral equations involving the Mittag–Leffler function can be obtained:

$$\int_0^\infty e^{-t}t^{\mu-1}E_{\lambda,\mu}\left(zt^\lambda\right)dt = \frac{1}{1-z}, \qquad (|z|<1). \tag{1.12}$$

The computation of the derivative and integration for the Mittag–Leffler function will be presented in the next section after the unique fractional differentiation and fractional integration are defined.

1.1.4   OTHER USEFUL FUNCTIONS

The Wright function is similar to the Mittag–Leffler function and it is defined as

$$W(z; \lambda, \mu) = \sum_{k=0}^{\infty} \frac{z^k}{k! \Gamma(\lambda k + \mu)}. \tag{1.13}$$

The Laplace transform of the Wright function is also given:

$$\mathcal{L}\{W(t; \lambda, \mu)\} = \sum_{k=0}^{\infty} \int_0^{\infty} e^{-st} \frac{t^k}{k! \Gamma(\lambda k + \mu)} dt$$

$$= \sum_{k=0}^{\infty} \frac{1}{\Gamma(\lambda k + \mu)} \cdot \frac{1}{s^{k+1}} = s^{-1} E_{\lambda,\mu}(s^{-1}). \tag{1.14}$$

The error function is frequently used in the calculus theory so that we give its definition in Equation (1.15). Correspondingly, the complementary error function is also given in Equation (1.16):

$$erf(z) = \frac{1}{\sqrt{\pi}} \int_{-z}^{z} e^{-t^2} dt = \frac{2}{\sqrt{\pi}} \int_0^z e^{-t^2} dt \tag{1.15}$$

$$erfc(z) = 1 - erf(z) = \frac{2}{\sqrt{\pi}} \int_z^{\infty} e^{-t^2} dt. \tag{1.16}$$

The definition of the Dawson function is given in Equation (1.17) and a known property for the Dawson function is given in Equation (1.18):

$$daw(z) = e^{-z^2} \int_0^z e^{t^2} dt \tag{1.17}$$

$$\frac{d}{dz} daw(z) = 1 - 2z \cdot daw(z). \tag{1.18}$$

Some other functions are also useful in the applications of the fractional calculus in the field of engineering and technology. These functions include Agarwal Function, Erdelyi's function, Robotnov–Hartley function, Miller–Ross function, and hypergeometric function [Das, 2008, Magin, 2006, Petás, 2011]. The complete introduction to these functions are omitted and the applications of them will be seen in the following chapters.

1.2   DEFINITIONS OF FRACTIONAL DERIVATIVES AND INTEGRALS

In this section, several definitions of the fractional derivatives and integrals, namely Grünwald–Letnikov definition, Riemann–Liouville definition, Caputo's definition, and Milller–Ross definition, are introduced. For the case of zero initial states, the Grünwald–Letnikov definition and

Riemann–Liouville definition can be employed to model the fractional-order dynamic plant. On the other hand, the definition proposed by Caputo in 1967 is more useful in the applications with non-zero initial states. Milller–Ross derivative is a special case of the sequential definition. The sequential definition is essentially a generalization of the above mentioned four definitions.

In the beginning of this section, Cauchy integrals from the integer-order calculus and the fractional-order calculus are given. Equation (1.19) is Cauchy integral for the integer-order integration, while Equation (1.20) is Cauchy integral for the arbitrary real-order integration. And Equations (1.21) and (1.22) are Cauchy integrals for the integer-order derivative and the arbitrary real-order derivative, respectively. In these equations, n is the positive integer, and p is the positive real number:

$$D^{-n} f(t) = \frac{1}{(n-1)!} \int_a^t (t-\tau)^{n-1} f(\tau) \, d\tau \tag{1.19}$$

$$D^{-p} f(t) = \frac{1}{\Gamma(p)} \int_a^t (t-\tau)^{p-1} f(\tau) \, d\tau \tag{1.20}$$

$$D^n f(t) = \frac{n!}{2\pi j} \oint \frac{f(\tau)}{(\tau-t)^{n+1}} \, d\tau \tag{1.21}$$

$$D^p f(t) = \frac{\Gamma(p+1)}{2\pi j} \oint \frac{f(\tau)}{(\tau-t)^{p+1}} \, d\tau. \tag{1.22}$$

1.2.1 GRÜNWALD–LETNIKOV FRACTIONAL DERIVATIVES

The Grünwald–Letnikov fractional calculus is defined as Equation (1.23); note that $\alpha > 0$ if it is differentiation and $\alpha < 0$ if it is integration:

$$\begin{aligned}
{}^{GL}_a D_t^\alpha f(t) &= \lim_{h \to 0} \frac{1}{h^\alpha} \sum_{i=0}^{[(t-a)/h]} (-1)^i \binom{\alpha}{i} f(t-ih) \\
&= \lim_{h \to 0} \frac{1}{h^\alpha} \sum_{i=0}^{[(t-a)/h]} c_i f(t-ih).
\end{aligned} \tag{1.23}$$

In Equation (1.23), the combinatorics $\binom{\alpha}{i}$ is evaluated as follows:

$$\binom{\alpha}{i} = \frac{\alpha(\alpha-1)\cdots(\alpha-i+1)}{i(i-1)\cdots 1} = \frac{\Gamma(\alpha+1)}{\Gamma(i+1)\Gamma(\alpha-i+1)}. \tag{1.24}$$

It is proven that the limit in Equation (1.23) can be evaluated under the conditions such that $m < \alpha < m+1$ and m is a positive integer. With the assumption that the derivatives $f^{(k)}(t)$, $(k = 1, 2, \ldots, m+1)$ are continuous in $[a, t]$, i.e., the function $f(t)$ is at least

$(m + 1)$-th order continuously differentiable, the following formula can be derived:

$$
{}^{GL}_{a}D^{\alpha}_{t} f (t) = \sum_{k=0}^{m} \frac{f^{(k)} (a) (t - a)^{-\alpha+k}}{\Gamma(-\alpha + k + 1)}
$$

$$
+ \frac{1}{\Gamma(-\alpha + m + 1)} \int_{a}^{t} (t - \tau)^{m-\alpha} f^{(m+1)} (\tau) \, d\tau. \tag{1.25}
$$

Now it is quite straightforward to calculate the Grünwald–Letnikov fractional derivative of the power function. It is not cumbersome to show that Equation (1.26) holds true both for two cases: (1) $\alpha < 0, \beta > -1$; (2) $0 \leq m \leq \alpha < m + 1, \beta > m$:

$$
{}^{GL}_{a}D^{\alpha}_{t} (t - a)^{\beta} = \frac{\Gamma (\beta + 1)}{\Gamma (-\alpha + \beta + 1)} (t - a)^{\beta-\alpha} . \tag{1.26}
$$

The composition rule for the Grünwald–Letnikov fractional derivative is shown in the next part, with which the operations that involve the calculation of multi-folded Grünwald–Letnikov fractional derivative can be simplified.

However, applying these composition rules needs some prerequisites. The composition rules exist not only for the Grünwald–Letnikov fractional derivative, but also for the other definitions of the fractional derivatives which will be introduced later:

$$
\frac{d^{n}}{dt^{n}} \left({}^{GL}_{a}D^{\alpha}_{t} f (t) \right) = {}^{GL}_{a}D^{\alpha}_{t} \left(\frac{d^{n} f(t)}{dt^{n}} \right) + \sum_{k=0}^{n-1} \frac{f^{(k)} (a) (t - a)^{-\alpha-n+k}}{\Gamma(-\alpha - n + k + 1)}. \tag{1.27}
$$

Equation (1.27) is the first composition rule for the Grünwald–Letnikov fractional derivative; it is found that the compositional computation of an integer-order derivative and a fractional-order Grünwald–Letnikov fractional derivative is commutative only if $f^{(k)} (a) = 0$ holds true for $k = 0, 1, 2, \ldots, n - 1$.

For the second composition rule of the Grünwald–Letnikov fractional derivatives,

$$
{}^{GL}_{a}D^{\beta}_{t} \left({}^{GL}_{a}D^{\alpha}_{t} f (t) \right) = {}^{GL}_{a}D^{\alpha+\beta}_{t} f(t). \tag{1.28}
$$

The prerequisites are a little more complicated than the case for the first composition rule. Its prerequisites can be categorized into two cases.

When $\alpha < 0$, the composition rule in Equation (1.28) holds true for any real β value. When $0 \leq m < \alpha < m + 1$, the composition rule in Equation (1.28) holds true with

$$
f^{(k)} (a) = 0, \quad (k = 0, 1, \ldots, m - 1).
$$

Given Equation (1.28) and two prerequisites stated above, a more powerful commutative composition rule for the Grünwald–Letnikov fractional derivative can be generalized as follows:

$$
{}^{GL}_{a}D^{\beta}_{t} \left({}^{GL}_{a}D^{\alpha}_{t} f (t) \right) = {}^{GL}_{a}D^{\alpha+\beta}_{t} f(t) = {}^{GL}_{a}D^{\alpha}_{t} \left({}^{GL}_{a}D^{\beta}_{t} f (t) \right). \tag{1.29}
$$

Consider the case where $0 \leq m < \alpha < m + 1$ and $0 \leq n < q < n + 1$, we can define $r = \max(m, n)$. It is shown that the commutative composition rule in Equation (1.29) holds true if $f^{(k)} = 0$ for $k = 0, 1, \ldots, r - 1$.

1.2.2   RIEMANN–LIOUVILLE FRACTIONAL DERIVATIVES

Riemann–Liouville fractional derivative of the objective function $f(t)$ is equivalent to its Grünwald–Letnikov fractional derivative if it is $(m + 1)$-th order continuously differentiable. The definition of the Riemann–Liouville fractional derivative is given in Equation (1.30), where $m \leq \alpha < m + 1$:

$$^{RL}_{a}D_t^{\alpha} f(t) = \frac{1}{\Gamma(m + 1 - \alpha)} \left(\frac{d}{dt} \right)^{m+1} \int_a^t \frac{f(\tau)}{(t - \tau)^{\alpha - m}} \, d\tau. \tag{1.30}$$

It is very simple to verify the equivalence between the Riemann–Liouville definition and the Grünwald–Letnikov fractional derivative, as we perform the integration by parts to Equation (1.30) repeatedly, that is,

$$^{RL}_{a}D_t^{\alpha} f(t) = \frac{1}{\Gamma(m + 1 - \alpha)} \left(\frac{d}{dt} \right)^{m+1}$$
$$\left\{ \left[-\frac{1}{m - \alpha + 1} f(\tau)(t - \tau)^{m - \alpha + 1} \right] \Big|_a^t + \int_a^t \frac{(t - \tau)^{m - \alpha + 1}}{m - \alpha + 1} f'(\tau) \, d\tau \right\}. \tag{1.31}$$

Repeat this procedure m times, the iterative expression is obtained:

$$^{RL}_{a}D_t^{\alpha} = \sum_{k=0}^{m} \frac{f^{(k)}(a)(t - a)^{-\alpha + k}}{\Gamma(-\alpha + k + 1)} + \frac{1}{\Gamma(-\alpha + m + 1)} \int_a^t (t - \tau)^{m - \alpha} f^{(m+1)}(\tau) \, d\tau$$
$$= {}^{GL}_{a}D_t^{\alpha}. \tag{1.32}$$

The Riemann–Liouville fractional derivative also holds the similar composition rules. We will prove that the composition rule of Riemann–Liouville fractional derivative is essentially equivalent to that of the Grünwald–Letnikov fractional derivative. First, let's state the composition rules of the Riemann–Liouville fractional derivatives as follows. The proof was given in Podlubny [1999]:

$$\frac{d^n}{dt^n} \left({}^{RL}_{a}D_t^{\alpha} f(t) \right) = {}^{RL}_{a}D_t^{\alpha + n} f(t). \tag{1.33}$$

Equations (1.33) and (1.34) are the mixed integer-order and fractional-order composition rule for the Riemann–Liouville fractional derivative. Obviously, Equations (1.33) and (1.34) of Riemann–Liouville fractional derivative are identical to Equation (1.27) of Grünwald–Letnikov fractional derivative:

$$^{RL}_{a}D_t^{\alpha} \left(\frac{d^n f(t)}{dt^n} \right) = {}^{RL}_{a}D_t^{\alpha + n} f(t) - \sum_{k=0}^{n-1} \frac{f^{(k)}(a)(t - a)^{k - \alpha - n}}{\Gamma(1 + k - \alpha - n)}. \tag{1.34}$$

Next, the rules to calculate the composition of two fractional Riemann–Liouville operators are presented as follows in Equations (1.35) and (1.36), where $0 \le m \le \alpha < m + 1$ and $0 \le n \le \beta < n + 1$:

$$
{}^{RL}_a D_t^{\beta} \left({}^{RL}_a D_t^{\alpha} f(t) \right) = {}^{RL}_a D_t^{\alpha+\beta} f(t) - \sum_{k=1}^{m+1} \left[{}^{RL}_a D_t^{\alpha-k} f(t) \right]_{t=a} \frac{(t-a)^{-\beta-k}}{\Gamma(1-\beta-k)}. \tag{1.35}
$$

$$
{}^{RL}_a D_t^{\alpha} \left({}^{RL}_a D_t^{\beta} f(t) \right) = {}^{RL}_a D_t^{\alpha+\beta} f(t) - \sum_{k=1}^{n+1} \left[{}^{RL}_a D_t^{\beta-k} f(t) \right]_{t=a} \frac{(t-a)^{-\alpha-k}}{\Gamma(1-\alpha-k)}. \tag{1.36}
$$

Equations (1.37) and (1.38) demonstrate the composition rule between the fractional Riemann–Liouville integration and fractional Riemann–Liouville derivative, where $0 \le l \le \gamma < l + 1$. It is seen from Equation (1.37) that the sum of the fractional integrals evaluated at $t = a$ is 0 since $-\gamma - k$ is always negative and $\left[{}^{RL}_a D_t^{-\gamma-k} f(t) \right]_{t=a} = 0$:

$$
{}^{RL}_a D_t^{\alpha} \left({}^{RL}_a D_t^{-\gamma} f(t) \right) = {}^{RL}_a D_t^{\alpha-\gamma} f(t). \tag{1.37}
$$

$$
{}^{RL}_a D_t^{-\gamma} \left({}^{RL}_a D_t^{\alpha} f(t) \right) = {}^{RL}_a D_t^{\alpha-\gamma} f(t) - \sum_{k=1}^{l+1} \left[{}^{RL}_a D_t^{\alpha-k} f(t) \right]_{t=a} \frac{(t-a)^{\gamma-k}}{\Gamma(1+\gamma-k)}. \tag{1.38}
$$

Therefore, it is concluded that Equations (1.35) and (1.36) can be universally applied to the composition of any two types of Riemann–Liouville fractional operators. For example, ${}^{RL}_a D_t^{\alpha} \left({}^{RL}_a D_t^{\beta} f(t) \right)$, ${}^{RL}_a D_t^{\beta} \left({}^{RL}_a D_t^{\alpha} f(t) \right)$, ${}^{RL}_a D_t^{-\alpha} \left({}^{RL}_a D_t^{\beta} f(t) \right)$, or ${}^{RL}_a D_t^{\alpha} \left({}^{RL}_a D_t^{-\beta} f(t) \right)$ for $\alpha, \beta > 0$.

If we assume function $f(t)$ is m-times continuously differentiable, and its m-th derivative is integrable in $[a, \infty]$, it can be verified that the following two conditions are equivalent. This equivalence can be easily observed from Equation (1.25).

$$
\left[{}^{RL}_a D_t^{\alpha} f(t) \right]\big|_{t=a} = 0 \tag{1.39}
$$

$$
f^{(k)}(a) = 0, \quad \text{for} \quad k = 0, 1, 2, \ldots, m. \tag{1.40}
$$

From Equations (1.35) and (1.36), the composition of two Riemann–Liouville fractional operators is actually not commutative. However, they are commutative if we have the following two conditions in Equations (1.41) and (1.42):

$$
\left[{}^{RL}_a D_t^{\alpha-k} f(t) \right]\bigg|_{t=a} = 0, \quad \text{for} \quad k = 1, 2, \ldots, m + 1. \tag{1.41}
$$

$$
\left[{}^{RL}_a D_t^{\beta-k} f(t) \right]\bigg|_{t=a} = 0, \quad \text{for} \quad k = 1, 2, \ldots, n + 1. \tag{1.42}
$$

Due to the equivalence of Equation (1.39) and Equation (1.40), Equations (1.41) and (1.43) are also equivalent. If the condition $f^{(k)}(a) = 0$, $(k = 0, 1, \ldots, r - 1)$, $r = \max(m, n)$ is satisfied, Equations (1.42) and (1.43) are satisfied, and the composition of two Riemann–Liouville operators are commutative. This condition is identical to the one for the composition of two Grünwald–Letnikov fractional operators, which is stated in Equation (1.29) since Grünwald–Letnikov fractional definition is essentially equivalent to the Riemann–Liouville fractional definition under certain conditions:

$$f^{(k)}(a) = 0, \quad (k = 0, 1, \ldots, m - 1) \tag{1.43}$$

$$f^{(k)}(a) = 0, \quad (k = 0, 1, \ldots, n - 1). \tag{1.44}$$

1.2.3   CAPUTO'S FRACTIONAL DERIVATIVES

For the cases where Grünwald–Letnikov fractional definitions or Riemann–Liouville fractional definitions are adopted, the initial condition $\left[{}^{RL}_{a}D^{\alpha}_t \right]_{t=a}$ is difficult to be accurately measured with sensors, and this limitation results in the rare usage of Grünwald–Letnikov and Riemann–Liouville fractional operators in real fractional-order system modeling, since most systems have non-zero initial conditions. Alternatively, Caputo's fractional derivative is defined and given in Equation (1.45), where $0 \leq m \leq \alpha < m + 1$, and it is developed as a useful tool to tackle the non-zero initial value problems:

$$ {}^{C}_{a}D^{\alpha}_t f(t) = \frac{1}{\Gamma(m + 1 - \alpha)} \int_a^t \frac{f^{(m+1)}(\tau)}{(t - \tau)^{\alpha - m}} \, d\tau. \tag{1.45}$$

From Equation (1.25), it can be derived that Caputo's fractional operator and the Riemann–Liouville or Grünwald–Letnikov fractional operator have a relation as follows:

$$ {}^{RL}_{a}D^{\alpha}_t f(t) = \sum_{k=0}^{m} \frac{f^{(k)}(a)(t - a)^{-\alpha + k}}{\Gamma(-\alpha + k + 1)} + {}^{C}_{a}D^{\alpha}_t f(t). \tag{1.46}$$

By observing Equations (1.45) and (1.46), Riemann–Liouville, Grünwald–Letnikov, and Caputo's fractional derivatives can be treated as the interpolation between two consecutive integer-order derivatives. We can see this with the proof as follows.

Proof. First, let us expand the definition of Caputo's derivative by assuming $f(t)$ has $(m + 2)$-times continuous and bounded derivatives in $[a, \infty]$. The first term of Equation (1.45) will be cancelled by applying the useful property of the gamma function: $\frac{1}{\Gamma(-n)} = 0$ for $n = 0, 1, 2, \ldots$

Finding the limit of Caputo's fractional derivative as its order tends to an integer, this can be done with integration by parts:

$$\lim_{\alpha \to m+1} {}_a^C D_t^\alpha f(t)$$

$$= \lim_{\alpha \to m+1} \left(\frac{f^{(k)}(a)(t-a)^{-\alpha+k}}{\Gamma(m-\alpha+2)} + \frac{1}{\Gamma(m-\alpha+2)} \int_a^t (t-\tau)^{m+1-\alpha} f^{(m+2)}(\tau) d\tau \right)$$

$$= f^{(m+1)}(a) + \int_a^t f^{(m+2)}(\tau) d\tau = f^{(m+1)}(t), \ (m = 0, 1, 2, \ldots).$$

Similarly, taking limit on both sides of Riemann–Liouville fractional derivative, gives

$$\lim_{\alpha \to m+1} {}^{RL}_a D_t^\alpha f(t) = \lim_{\alpha \to m+1} \sum_{k=0}^m \frac{f^{(k)}(a)(t-a)^{-\alpha+k}}{\Gamma(-\alpha+k+1)} + \lim_{\alpha \to m+1} {}^C_a D_t^\alpha f(t)$$

$$= \sum_{k=0}^m \frac{f^{(k)}(a)(t-a)^{-m+k-1}}{\Gamma(-m+k)} + f^{(m+1)}(t) = f^{(m+1)}(t).$$

$$\square$$

Resulted from the interpolation property, Caputo's fractional derivative of a constant is always 0. However, the Riemann–Liouville or Grünwald–Letnikov fractional derivative of a constant number is not zero by observing Equation (1.45), and its value is given by

$$ {}^{RL}_a D_t^\alpha C = {}^{GL}_a D_t^\alpha C = \frac{C(t-a)^{-\alpha}}{\Gamma(1-\alpha)}. \tag{1.47}$$

Correspondingly, the computation rule of composing an integer-order derivative and a fractional-order derivative can be obtained from Equations (1.33), (1.34), and (1.45), which is given below, and satisfied when $f^{(k)}(t) = 0$, $k = m+1, m+2, \ldots, n-1$:

$$ {}^C_a D_t^\alpha \left(\frac{d^n}{dt^n} f(t) \right) = \frac{d^n}{dt^n} \left({}^C_a D_t^\alpha f(t) \right) = {}^C_a D_t^{\alpha+n} f(t). \tag{1.48}$$

The computation rule of nesting one Caputo's fractional-order derivative in another fractional-order derivative can be derived in an analogous way by considering Equations (1.35), (1.36), and (1.45).

Caputo's fractional definition is popular in real system modeling, especially for fractional systems with nonzero initial states. The fractional differential equations defined with Caputo's fractional operators have integer-order initial states $f^{(i)}(0)$, which can be seen from the Laplace transform of Caputo's derivative in Equation (1.62). The integer-order initial states are easier to be measured from sensors.

1.2.4 MILLER–ROSS FRACTIONAL DERIVATIVES

As stated in integer-order calculus theory, the n-th order derivative can be expressed as a sequence of first-order derivative operators. Analogously, Miller–Ross fractional derivative operator is defined as a sequence of Riemann–Liouville fractional derivative operators, i.e.,

$$
{}^{MR}_{a}D^{\alpha}_{t} f(t) = {}^{RL}_{a}D^{\alpha_1}_{t} {}^{RL}_{a}D^{\alpha_2}_{t} \cdots {}^{RL}_{a}D^{\alpha_n}_{t} f(t), \tag{1.49}
$$

where $\alpha = \alpha_1 + \alpha_2 + \cdots + \alpha_n$ and $0 \le \alpha_i \le 1$ for $i = 1, \ldots n$ in Equation (1.49).

Sequential fractional derivative is a more general form of the Miller–Ross fractional derivative. In the definition for a general-form sequential fractional derivative, the Riemann–Liouville operators in Miller–Ross sequence can be replaced by Grünwald–Letnikov fractional operators or Caputo's fractional operators. Hence, more different types of fractional operators can be defined by manipulating the orders or types of components in sequential fractional derivative.

Evidently, the Riemann–Liouville fractional derivative can be expressed in a sequential form as Equation (1.50), where $n \le \alpha < n + 1$:

$$
{}^{RL}_{a}D^{\alpha}_{t} f(t) = \underbrace{\frac{d}{dt}\frac{d}{dt} \cdots \frac{d}{dt}}_{n+1} {}^{RL}_{a}D^{-(n+1-\alpha)}_{t} f(t). \tag{1.50}
$$

The Caputo's fractional derivative can also be written in a sequential form:

$$
{}^{C}_{a}D^{\alpha}_{t} f(t) = {}^{RL}_{a}D^{-(n+1-\alpha)}_{t} \underbrace{\frac{d}{dt}\frac{d}{dt} \cdots \frac{d}{dt}}_{n+1} f(t). \tag{1.51}
$$

As discussed above, any type of fractional derivative operators can be expressed as an equivalent Miller–Ross form composed of only sequential integer-order derivative operators and Riemann–Liouville fractional operators. This uniquely defined Miller–Ross representation implies that different types of fractional operators may hold some properties in common. In summary to what we discussed above, we will give the proof of some common properties of fractional operators in the following subsection.

1.2.5 PROPERTIES OF FRACTIONAL OPERATORS

The fractional operators have two types: the fractional derivative operators and the fractional integral operators. Both types share several common properties which are widely used in the applications of fractional calculus theory. These properties not only hold for the Grünwald–Letnikov or the Riemann–Liouville fractional definitions, but also for the Caputo's or Miller–Ross fractional definitions. Moreover, these properties are widely used to simplify the computations for solving the fractional-order differential equations.

Linearity: the linearity of the fractional operators is shown in Equation (1.52), where λ_1 and λ_2 are constants:

$$D^\alpha \left(\lambda_1 f(t) + \lambda_2 g(t) \right) = \lambda_1 D^\alpha f(t) + \lambda_2 D^\alpha g(t). \tag{1.52}$$

Proof of the linearity of the Caputo's fractional operator. Assume that $m \le \alpha < m + 1$.

$$
\begin{aligned}
{}_a^C D_t^\alpha \left(\lambda_1 f(t) + \lambda_2 g(t) \right) &= \frac{1}{\Gamma(m+1-\alpha)} \int_a^t \frac{\lambda_1 f^{(m+1)}(\tau) + \lambda_2 g^{(m+1)}(\tau)}{(t-\tau)^{\alpha-m}} \, d\tau \\
&= \frac{\lambda_1}{\Gamma(m+1-\alpha)} \int_a^t \frac{f^{(m+1)}(\tau)}{(t-\tau)^{\alpha-m}} \, d\tau \\
&\quad + \frac{\lambda_2}{\Gamma(m+1-\alpha)} \int_a^t \frac{g^{(m+1)}(\tau)}{(t-\tau)^{\alpha-m}} \, d\tau \\
&= \lambda_1 {}_a^C D_t^\alpha f(t) + \lambda_2 {}_a^C D_t^\alpha g(t).
\end{aligned}
$$

$\square$

For the fractional operators with other definitions, the proof is similar.

Leibniz rule: the Leibniz rule is a decent result from integer-order calculus, and it is stated in Equation (1.53). Here, we will give the proof that the Leibniz rule also holds true for the fractional-order calculus, as shown in Equation (1.54), where $\alpha \ne -1, -2, \ldots$ and $t \ne 0$:

$$\frac{d^n}{dt^n} \left(f(t)g(t) \right) = \sum_{i=0}^n \binom{n}{i} f^{(i)}(t) g^{(n-i)}(t). \tag{1.53}$$

$$D^\alpha \left(f(t)g(t) \right) = \sum_{i=0}^\infty \binom{\alpha}{i} f^{(i)}(t) D^{\alpha-i} g(t). \tag{1.54}$$

Proof of Leibniz rule in Equation (1.54). First, let us expand function $f(\tau)$ as Taylor series at $\tau = t$, that is,

$$f(\tau) = \sum_{i=0}^\infty \frac{f^{(i)}(t)(\tau - t)^i}{i!}. \tag{1.55}$$

If we multiply both sides of Equation (1.55) by $\frac{\Gamma(1+\alpha)(\tau-t)^{-\alpha-1} g(\tau)}{2\pi j}$, and integrate along the same contour of the Cauchy integral in Equation (1.22), we have:

$$\frac{\Gamma(\alpha+1)}{2\pi j} \oint \frac{f(\tau) g(\tau)}{(\tau-t)^{\alpha+1}} \, d\tau = \frac{\Gamma(1+\alpha)}{2\pi j} \frac{\Gamma(1+\alpha-i)}{\Gamma(1+\alpha-i)} \sum_{i=0}^\infty \oint \frac{g(\tau) \, d\tau}{i!\,(\tau-t)^{\alpha+1-i}} f^{(i)}(t). \tag{1.56}$$

By comparing Equations (1.22) and (1.56), we get:

$$D^{\alpha}\left(f\left(t\right)g\left(t\right)\right) = \frac{\Gamma\left(1+\alpha\right)}{\Gamma\left(1+\alpha-i\right)i!} \sum_{i=0}^{\infty} \frac{\Gamma\left(1+\alpha-i\right)}{2\pi j} \oint \frac{g\left(\tau\right)d\tau}{\left(\tau-t\right)^{\alpha+1-i}} f^{(i)}\left(t\right)$$

$$= \sum_{i=0}^{\infty} \frac{\Gamma(1+\alpha)}{\Gamma(1+\alpha-i)i!} D^{\alpha-i}g(t)f^{(i)}(t) \qquad (1.57)$$

$$= \sum_{i=0}^{\infty} \binom{\alpha}{i} f^{(i)}(t)D^{\alpha-i}g(t).$$

$\square$

It should be noted that there is an alternative way to prove the Leibniz rule, which was presented in Podlubny [1999]. In this book, the well-known Cauchy integral, i.e., Equation (1.22), are utilized in the proof, which greatly simplifies its proof steps.

1.3   FRACTIONAL DIFFERENTIAL EQUATIONS

The theory of the ordinary differential equations in integer-order calculus was fully developed. And as an extension, the fractional-order ordinary differential equations are studied in this section. We first discuss the existence and uniqueness conditions for the solution of a fractional differential equation. Provided that the fractional differential equation is linear with constant coefficients, the exact solution for the equation can be solved analytically with various methods. For example, the Laplace transform method, the Mellin transform method and the fractional Green's function. In the beginning subsections, these transformation methods are introduced one by one, and in the last section, we focus on how to how to numerically solve the fractional differential equation.

1.3.1   LINEAR AND NONLINEAR FRACTIONAL DIFFERENTIAL EQUATIONS

As it is discussed in the beginning of the last section, if a different definition of the fractional derivatives are chosen, a different type of fractional differential equations is given. For our case, since the systems most widely studied are those with zero initial states, the Riemann–Liouville's definition will be appropriate to define the fractional differential equations in this book.

The simplest type of fractional differential equations is the linear fractional ordinary differential equation with constant coefficients, which is used to model the linear time invariant (LTI) system. The linear fractional differential equation with constant coefficients has a general form which is given in Equation (1.58), where we assume the index is placed in descending order: $\alpha_1 > \alpha_2 > \cdots > \alpha_{n-1} > \alpha_n$ and $\beta_1 > \beta_2 > \cdots > \beta_{m-1} > \beta_m$:

$$a_1 \, {}^{RL}_a D_t^{\alpha_1} y(t) + a_2 \, {}^{RL}_a D_t^{\alpha_2} y(t) + \cdots + a_n \, {}^{RL}_a D_t^{\alpha_n} y(t) = b_1 \, {}^{RL}_a D_t^{\beta_1} u(t) + b_2 \, {}^{RL}_a D_t^{\beta_2} u(t) + \cdots + b_m \, {}^{RL}_a D_t^{\beta_m} u(t). \tag{1.58}$$

With the aim to describe the fractional-order LTI system using its transfer function, it is necessary to introduce the Laplace transform of the Riemann–Liouville fractional derivative. The Laplace transform of the integer-order derivative and the Caputo's fractional derivative is given as a reference as follows.

The Laplace transform of the integer-order derivative is shown in Equation (1.59):

$$\mathcal{L}\left\{\frac{d^n}{dt^n} f(t)\right\} = s^n F(s) - \sum_{i=1}^{n} s^{n-i} f^{(i-1)}(0). \tag{1.59}$$

The Laplace transform of the Riemann–Liouville fractional derivative is shown in Equation (1.60), where $n < \alpha \le n + 1$:

$$\mathcal{L}\left\{{}^{RL}_a D_t^{\alpha} f(t)\right\} = s^{\alpha} F(s) - \sum_{i=0}^{n} s^{i} \, {}^{RL}_a D_t^{\alpha-i-1} f(0). \tag{1.60}$$

Proof of Equation (1.60). Initially, let us give the proof of Equation (1.61):

$$\mathcal{L}\left\{{}^{RL}_a D_t^{-\beta} f(t)\right\} = s^{-\beta} F(s), \quad (\beta > 0) \tag{1.61}$$

$${}^{RL}_a D_t^{-\beta} f(t) = \frac{1}{\Gamma(\beta)} \int_0^t \frac{f(\tau)}{(t-\tau)^{1-\beta}} \, d\tau = \frac{1}{\Gamma(\beta)} t^{\beta-1} * f(t).$$

The Laplace transform of $t^{\beta-1}$ is given below:

$$\mathcal{L}\left\{t^{\beta-1}\right\} = \int_0^{\infty} e^{-st} t^{\beta-1} dt = s^{-\beta} \int_0^{\infty} e^{-st} (st)^{\beta-1} d(st) = s^{-\beta} \Gamma(\beta).$$

Therefore, we have the following derivation according to the convolution theorem:

$$\mathcal{L}\left\{{}^{RL}_a D_t^{-\beta} f(t)\right\} = \frac{1}{\Gamma(\beta)} \mathcal{L}\left\{t^{\beta-1}\right\} \cdot \mathcal{L}\{f(t)\} = s^{-\beta} F(s).$$

Then, we are able to get the Laplace transform of the Riemann–Liouville derivative:

$$\mathcal{L}\left\{{}^{RL}_{a}D^{\alpha}_{t}f(t)\right\} = \int_{0}^{\infty} e^{-st} \frac{d^{n+1}}{dt^{n+1}} {}^{RL}_{a}D^{-(n+1-\alpha)}_{t}f(t)dt$$

$$= s^{(n+1)}\mathcal{L}\left\{{}^{RL}_{a}D^{-(n+1-\alpha)}_{t}f(t)\right\} - \sum_{i=1}^{n+1} s^{n+1-i}{}^{RL}_{a}D^{-(n+1-\alpha)+(i-1)}_{t}f(0)$$

$$= s^{\alpha}F(s) - \sum_{k=0}^{n} s^{n+1-(k+1)}{}^{RL}_{a}D^{-(n+1-\alpha)+(k+1-1)}_{t}f(0)$$

$$= s^{\alpha}F(s) - \sum_{j=0}^{n} s^{j}{}^{RL}_{a}D^{\alpha-1-j}_{t}f(0).$$

$\square$

The Laplace transform of the Caputo's fractional derivative is given in Equation (1.62) and its proof is presented as follows:

$$\mathcal{L}\left\{{}^{C}_{a}D^{\alpha}_{t}f(t)\right\} = s^{\alpha}F(s) - \sum_{i=0}^{n} s^{\alpha-i-1}f^{(i)}(0). \qquad (1.62)$$

Proof of Equation (1.62).

$$\mathcal{L}\left\{{}^{C}_{a}D^{\alpha}_{t}f(t)\right\} = \mathcal{L}\left\{{}^{C}_{a}D^{-(n+1-\alpha)}_{t} \frac{d^{n+1}}{dt^{n+1}}f(t)\right\}$$

$$= s^{-(n+1-\alpha)}\mathcal{L}\left\{\frac{d^{n+1}}{dt^{n+1}}f(t)\right\}$$

$$= s^{-(n+1-\alpha)}\left[s^{n+1}F(s) - \sum_{i=1}^{n+1} s^{n+1-i}f^{(i-1)}(0)\right]$$

$$= s^{\alpha}F(s) - \sum_{j=0}^{n} s^{\alpha-j-1}f^{(j)}(0).$$

$\square$

If the initial states are zeros, the equivalence between Equations (1.37) and (1.38) implies that the linear fractional differential equations defined with the Riemann–Liouville fractional derivatives as given in Equation (1.58), with the Caputo's fractional derivatives, or with the Grünwald–Letnikov fractional derivatives, have the same transfer function. The transfer function of the fractional-order LTI system can be obtained by applying Laplace transform to

Equation (1.58):

$$G(s) = \frac{Y(s)}{U(s)} = \frac{b_1 s^{\beta_1} + b_2 s^{\beta_2} + \cdots + b_m s^{\beta_m}}{a_1 s^{\alpha_1} + a_2 s^{\alpha_2} + \cdots + a_n s^{\alpha_n}}. \tag{1.63}$$

Compared to the fractional LTI system, the fractional nonlinear system does not have a transfer function with explicit form. Consider the general-form nonlinear fractional differential equation defined in Equation (1.64); analyzing its mathematical properties is a tough work. However, finding the approximate solutions to this equation with various numerical methods remains to be a hot research topic nowadays [Xue, 2017]:

$$F\left(t, y(t), {}^{RL}_a D^{\alpha_1}_t y(t), \ldots, {}^{RL}_a D^{\alpha_n}_t y(t)\right) = 0. \tag{1.64}$$

In Equation (1.64), $F(\cdot)$ is a highly nonlinear algebraic function. To solve the nonlinear fractional differential equation, pure mathematical techniques may lose their effectiveness. Therefore, the numerical methods are discussed at the end of this section to solve this type of highly nonlinear fractional differential equations.

1.3.2   ANALYSIS FOR EXISTENCE AND UNIQUENESS OF SOLUTIONS

In nonlinear control theory, the fixed point theorem gives the conditions for the existence and uniqueness of the solution to an integer-order nonlinear differential equation $\dot{x} = f(t, x)$, where x is state vector. The global existence and uniqueness requires $f(t, x)$ to have countably many discontinuities in t and be Lipshitz continuous in x, i.e., $\| f(t, x) - f(t, y) \| \leq L \| x - y \|$, where y is another state vector satisfying the integer-order nonlinear differential equation $\dot{y} = f(t, y)$. Without loss of generality, let us only consider the existence and uniqueness problem for the solution of Equation (1.65), where $0 < \alpha < 1$ and initial condition is ${}^{RL}_0 D^{\alpha-1}_t y(0) = c$:

$$ {}^{RL}_0 D^{\alpha}_t y(t) = f(t, y). \tag{1.65}$$

Let us define set G as Cartesian product of the time domain and state set $G = T \otimes X$. Given a closed subset $S(\delta, r) \subset G$ such that for any points $(t, y) \in S$, the following two inequalities hold true:

$$0 \leq t \leq \delta, \quad \left\| t^{1-\alpha} y(t) - \frac{c}{\Gamma(\alpha)} \right\| \leq r. \tag{1.66}$$

In order to prove the existence and uniqueness theorem, the contraction mapping theorem should be presented as a lemma, the proof for it is omitted and can be found in Khalil [2002] or Slotine and Li [1991].

Lemma 1.1   *(Contraction Mapping Theorem) For a closed subset S in a Banach space X, and a contraction mapping T that maps S into S, suppose that*

$$\| T(x) - T(y) \| \leq \rho \| x - y \|, \quad \forall x, y \in S, \quad 0 \leq \rho < 1.$$

Then, there exists a unique vector $x^* \in S$ satisfying $x^* = T(x^*)$. x^* can be solved by the successive approximation method, assigning initial value to any vector in S.

Provided with Lemma 1.1, it is possible for us to establish the existence and uniqueness theorem for the solution of the Equation (1.65) and condition (1.66).

Theorem 1.2   *(**Existence and Uniqueness**) Given $f(t, y)$ to be a real-valued continuous function with G as its domain, and $f(t, y)$ is Lipshitz continuous on y, i.e.,*

$$|f(t, x) - f(t, y)| \leq L|x - y|_C$$

and conditions in Equation (1.66) are satisfied. Then the problem in Equation (1.65) has a unique solution over the closed sub-domain $S \subset G$, where S is defined in Equation (1.66).

Proof of Theorem *1.2.*   First, by applying the definition for Riemann–Liouville fractional derivative, we can get two useful equations as below:

$$^{RL}_{0}D^{\alpha}_{t}\left(\frac{t^{\alpha-1}}{\Gamma(\alpha)}\right) = 0 \tag{1.67}$$

$$^{RL}_{0}D^{\alpha-1}_{t}\left(\frac{t^{\alpha-1}}{\Gamma(\alpha)}\right) = 1. \tag{1.68}$$

Then, let us consider a functional mapping P such that

$$y(t) = P(y(t)) := \frac{c}{\Gamma(\alpha)}t^{\alpha-1} + \frac{1}{\Gamma(\alpha)}\int_0^t (t - \tau)^{\alpha-1} f(\tau, y(\tau)) \, d\tau. \tag{1.69}$$

By applying Equation (1.67) and composition rule in Equation (1.37) to the both sides of Equation (1.69), it is easy to observe that Equation (1.69) is equivalent to the Equation (1.65), and if we apply Equation (1.68) to the both sides of Equation (1.69), we can get the initial conditions for the original fractional differential equation. Thus, it implies that the equation transformed from a mapping P is equivalent to the original fractional differential equation. Therefore, it is enough to show that the transformation P in Equation (1.69) is a contraction mapping which maps $f(t, y)$ from S to S.

Let us first show that the mapping P maps $f(t, y)$ from S to S. Suppose that conditions in Equation (1.66) hold true, i.e., $0 \leq t \leq \delta, \left|t^{1-\alpha}y(t) - \frac{c}{\Gamma(\alpha)}\right| \leq r$. Since function $f(t, y)$ is continuous and defined on G, and G is a compact set (closed and bounded), it implies that function $f(t, y)$ is bounded on G, say $|f(t, y)| \leq M$. Then, the following derivations are valid:

$$\left| t^{1-\alpha} P\left(y\left(t\right)\right) - \frac{c}{\Gamma\left(\alpha\right)} \right| = \left| \frac{t^{1-\alpha}}{\Gamma\left(\alpha\right)} \int_0^t (t - \tau)^{\alpha-1} f\left(\tau, y\left(\tau\right)\right) d\tau \right|$$

$$\leq \frac{t^{1-\alpha}}{\Gamma\left(\alpha\right)} \int_0^t (t - \tau)^{\alpha-1} |f\left(\tau, y\left(\tau\right)\right)| d\tau$$

$$\leq \frac{M t^{1-\alpha}}{\Gamma\left(\alpha\right)} \int_0^t (t - \tau)^{\alpha-1} d\tau$$

$$\leq \frac{M\delta}{\Gamma(\alpha + 1)}.$$

Pick $\delta \leq \frac{r\Gamma(\alpha+1)}{M}$, then it follows that $\left| t^{1-\alpha} P\left(y\left(t\right)\right) - \frac{c}{\Gamma(\alpha)} \right| \leq r$ and the mapping P maps $y(t)$ from S to S.

Next, let us give the proof that P is essentially a contraction mapping:

$$|P\left(x\left(t\right)\right) - P\left(y\left(t\right)\right)| = \frac{1}{\Gamma\left(\alpha\right)} \left| \int_0^t (t - \tau)^{\alpha-1} \left[f\left(\tau, x\left(\tau\right)\right) - f\left(\tau, y\left(\tau\right)\right) \right] d\tau \right|$$

$$\leq \frac{1}{\Gamma\left(\alpha\right)} \int_0^t (t - \tau)^{\alpha-1} |f\left(\tau, x\left(\tau\right)\right) - f\left(\tau, y\left(\tau\right)\right)| d\tau$$

$$\leq \frac{L}{\Gamma\left(\alpha\right)} \int_0^t (t - \tau)^{\alpha-1} |x - y|_C d\tau$$

$$\leq \frac{L\delta}{\Gamma\left(\alpha + 1\right)} |x - y|_C.$$

Choose that $\rho = \frac{L\delta}{\Gamma(\alpha+1)} < 1$, and pick $\delta = \min\left\{ \frac{\rho\Gamma(\alpha+1)}{L}, \frac{r\Gamma(\alpha+1)}{M} \right\}$, then we are able to conclude that P is a contraction mapping in the domain S. By Lemma 1.1 (Contraction Mapping Theorem), the proof of the existence and uniqueness theorem is completed. $\qquad\square$

In this section, the existence and uniqueness theorem is presented to discuss the existence and uniqueness conditions for the solution of a specific-type fractional differential equation. Even though this theorem cannot be treated as a general rule to promise the existence and uniqueness of the solution for any type of fractional differential equations, it is still useful and can be applied for many fractional order systems, whose system dynamics can be modeled to be a fractional differential equation, as shown in Equation (1.65).

1.3.3 ANALYTICAL SOLUTIONS TO LINEAR FRACTIONAL DIFFERENTIAL EQUATIONS

Examining the existence and uniqueness of the solution to the fractional order differential equation should be the first step of overall treatment for a fractional order system. In this subsequent section, three representative methods are presented to analytically solve the linear fractional differential equations.

1.3.3.1 The Laplace Transform Method

The Laplace transform method is a powerful tool to solve all types of linear differential equations, including the fractional linear differential equation and the integer-order linear differential equation. In Section 1.2, we have derived the Laplace transform for the Riemann–Liouville fractional derivative and the Caputo's fractional derivative in Equations (1.60) and (1.62). Besides that, we also render the Laplace transform of the Mittag–Leffler function in Equation (1.11), which can result in some useful formulas as follows:

$$\mathcal{L}^{-1}\left\{\frac{1}{s^{\lambda}-a}\right\}=t^{\lambda-1}E_{\lambda,\lambda}(at^{\lambda}) \tag{1.70}$$

$$\mathcal{L}^{-1}\left\{\frac{1}{s(s^{\lambda}-a)}\right\}=t^{\lambda}E_{\lambda,\lambda+1}(at^{\lambda}). \tag{1.71}$$

In Equation (1.70), we need to set $k=0$ and $\lambda=\mu$ in Equation (1.11); in Equation (1.71), we need to set $\lambda-\mu=-1$ and $k=0$. There is another important Laplace transform formula, which is useful for solving the linear commensurate fractional differential equations:

$$\mathcal{L}^{-1}\left\{\frac{s^{\lambda\alpha-\mu}}{(s^{\lambda}-a)^{\alpha}}\right\}=t^{\mu-1}E_{\lambda,\mu}^{\alpha}(at^{\lambda}). \tag{1.72}$$

By setting $\lambda\alpha-\mu=0$ and $\alpha=k$ as the integer, a corollary formula to the Equation (1.72) can be obtained in Equation (1.73). By setting $\lambda\alpha-\mu=-1$ and $\alpha=k$ as the integer, another corollary formula can be derived, as shown in Equation (1.74):

$$\mathcal{L}^{-1}\left\{\frac{1}{(s^{\lambda}-a)^{k}}\right\}=t^{\lambda k-1}E_{\lambda,\lambda k}^{k}(at^{\lambda}) \tag{1.73}$$

$$\mathcal{L}^{-1}\left\{\frac{1}{s(s^{\lambda}-a)^{k}}\right\}=t^{\lambda k}E_{\lambda,\lambda k+1}^{k}(at^{\lambda}). \tag{1.74}$$

Note $E_{\lambda,\mu}^{\alpha}(z)$ is the three-parameter Mittag–Leffler function and $E_{\lambda,\mu}^{1}(z)=E_{\lambda,\mu}(z)$. The introduction of the three-parameter Mittag–Leffler function is omitted in this book.

The following example illustrates how to apply the Laplace transform method to solve a linear commensurate fractional differential equation. Suppose the linear fractional differential equation is given in Equation (1.75) with zero initial state and the input signal is an ideal impulse signal,

$$\begin{aligned}&{}^{RL}_{0}D_{t}^{2.4}y\left(t\right)+10{}^{RL}_{0}D_{t}^{1.8}y\left(t\right)+35{}^{RL}_{0}D_{t}^{1.2}y\left(t\right)+50{}^{RL}_{0}D_{t}^{0.6}y\left(t\right)+24y\left(t\right)=\\&{}^{RL}_{0}D_{t}^{1.8}u\left(t\right)+6{}^{RL}_{0}D_{t}^{1.2}u\left(t\right)+10{}^{RL}_{0}D_{t}^{0.6}u\left(t\right)+2u(t).\end{aligned} \tag{1.75}$$

Taking Laplace transform on both sides of Equation (1.75), we can obtain the transfer function to this fractional LTI system, noting here we assume the initial states are zeros:

$$G(s) = \frac{Y(s)}{U(s)} = \frac{s^{1.8} + 6s^{1.2} + 10s^{0.6} + 2}{s^{2.4} + 10s^{1.8} + 35s^{1.2} + 50s^{0.6} + 24}$$

$$= \frac{1}{s^{0.6} + 4} - \frac{1}{2}\frac{1}{s^{0.6} + 3} + \frac{1}{s^{0.6} + 2} - \frac{1}{2}\frac{1}{s^{0.6} + 1}.$$

If the input signal is an ideal impulse signal, that means $\mathcal{L}\{u(t)\} = U(s) = 1$. Therefore, we can get the output signal of this fractional LTI system by applying inverse Laplace transform on the both sides of the transfer function. Using the inverse Laplace transform formula in Equation (1.73), we have

$$y(t) = \mathcal{L}^{-1}\{G(s)U(s)\}$$

$$= \mathcal{L}^{-1}\left\{\frac{1}{s^{0.6} + 4} - \frac{1}{2}\frac{1}{s^{0.6} + 3} + \frac{1}{s^{0.6} + 2} - \frac{1}{2}\frac{1}{s^{0.6} + 1}\right\}$$

$$= \mathcal{L}^{-1}\left\{\frac{1}{s^{0.6} + 4}\right\} - \frac{1}{2}\mathcal{L}^{-1}\left\{\frac{1}{s^{0.6} + 3}\right\} + \mathcal{L}^{-1}\left\{\frac{1}{s^{0.6} + 2}\right\} - \frac{1}{2}\mathcal{L}^{-1}\left\{\frac{1}{s^{0.6} + 1}\right\}$$

$$= t^{-0.4}E_{0.6,0.6}\left(v - 4t^{0.6}\right) - \frac{1}{2}t^{-0.4}E_{0.6,0.6}\left(-3t^{0.6}\right) + t^{-0.4}E_{0.6,0.6}\left(-2t^{0.6}\right)$$

$$- \frac{1}{2}t^{-0.4}E_{0.6,0.6}(-t^{0.6}).$$

As it is shown in the example above, Mittag–Leffler function plays an essential role in solving the linear fractional differential equations, more precisely speaking, the solution might be the linear combinations of several terms of Mittag–Leffler function. The Laplace transform of the Mittag–Leffler function in Equation (1.11) and the Laplace transform for the three-parameter Mittag–Leffler function in Equation (1.72) play the critical role, and many other corollary Laplace transform formulas are derived based on them. For examples, Equations (1.70), (1.71), (1.73), and (1.74) are the corollary formulas. The solution of the example in Equation (1.75) is plotted in Figure 1.3.

1.3.3.2 The Mellin Transform Method

The Mellin transform is defined in Equation (1.76), where $f(t)$ is a function on $(0, \infty)$ and z is the complex variable that is constricted in a vertical band zone, i.e., $\zeta_1 < Re(z) < \zeta_2$:

$$F(z) = \mathcal{M}\{f(t)\} = \int_0^\infty f(t)t^{z-1}dt. \tag{1.76}$$

The inverse Mellin transform is defined in Equation (1.77), where the integral contour is an infinite vertical line at real value equal to c. The Mellin transform pair exits when function

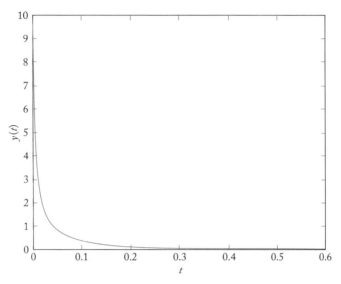

Figure 1.3: Solution to Equation (1.75) with impulse input.

$f(t)$ is continuous almost everywhere in any close subintervals in $(0, \infty)$:

$$f(t) = \mathcal{M}^{-1}\{F(z)\} = \frac{1}{2\pi j}\int_{\zeta - j\infty}^{\zeta + j\infty} F(z)\, t^{-z}\, dz. \qquad (1.77)$$

A useful property follows from the definition of the Mellin transform is given in Equation (1.78):

$$\mathcal{M}\{t^\gamma f(t)\} = \int_0^\infty f(t) t^{\gamma + z - 1}\, dt = F(z + \gamma). \qquad (1.78)$$

The Mellin convolution is defined in Equation (1.79), and the convolution theorem for the Mellin transform is shown in Equation (1.80) [EqWorld, 2019]:

$$f(t) * g(t) = \int_0^\infty f(t\tau)\, g(\tau)\, d\tau \qquad (1.79)$$

$$\mathcal{M}\{f(t) * g(t)\} = F(z)G(1 - z). \qquad (1.80)$$

Proof for the Equation (1.80). Fubini's theorem [Wheeden and Zygmund, 2015] can be applied in this proof,

$$\mathcal{M}\{f(t) * g(t)\} = \mathcal{M}\left\{\int_0^\infty f(t\tau)g(\tau)\,d\tau\right\}$$

$$= \int_0^\infty t^{z-1}\int_0^\infty f(t\tau)g(\tau)\,d\tau\,dt$$

$$= \int_0^\infty g(\tau)\int_0^\infty f(t\tau)t^{z-1}dt\,d\tau$$

$$= \int_0^\infty g(\tau)\int_0^\infty f(u)\left(\frac{u}{\tau}\right)^z\frac{du}{u}\,d\tau$$

$$= \int_0^\infty g(\tau)\tau^{-z}\,d\tau\int_0^\infty f(u)u^{z-1}\,du$$

$$= G(1-z)F(z).$$

□

Combining Equations (1.78) and (1.80) gives a new formula as follows:

$$\mathcal{M}\left\{t^\gamma\int_0^\infty \tau^\lambda f(t\tau)g(\tau)\,d\tau\right\} = F(z+\gamma)G(1-z-\gamma+\lambda). \tag{1.81}$$

The Mellin transform of the Riemann–Liouville fractional derivative is given in Equation (1.82); the Mellin transform of the Caputo's fractional derivative is given in Equation (1.83). The proofs of Equations (1.82) and (1.83) are given in Podlubny [1999]. Assume that $n < \alpha \le n + 1$, then we get:

$$\mathcal{M}\{{}^{RL}_0D_t^\alpha f(t)\} = \sum_{i=0}^n \frac{\Gamma(1-z+i)}{\Gamma(1-z)}\left[{}^{RL}_0D_t^{\alpha-i-1}f(t)t^{z-i-1}\right]_0^\infty$$

$$+ \frac{\Gamma(1-z+\alpha)}{\Gamma(1-z)}F(z-\alpha) \tag{1.82}$$

$$\mathcal{M}\{{}^C_0D_t^\alpha f(t)\} = \sum_{i=0}^n \frac{\Gamma(\alpha-z-n-1)}{\Gamma(1-z)}\left[f^{(i)}(t)t^{z-\alpha+i}\right]_0^\infty$$

$$+ \frac{\Gamma(1-z+\alpha)}{\Gamma(1-z)}F(z-\alpha). \tag{1.83}$$

If terms with the limits at ∞ and 0 are assumed to be zero both in Equations (1.82) and (1.83), then Equations (1.82) and (1.83) coincide. Under these circumstances, from Equation (1.78) a general formula as shown in Equation (1.84) is derived:

$$\mathcal{M}\left\{\sum_{l=0}^{N} a_l t^{\alpha+l} D^{\alpha+l} f(t)\right\} = F(z)\,\Gamma(1-z) \sum_{l=0}^{N} \frac{a_l}{\Gamma(1-z-\alpha-l)}. \tag{1.84}$$

With the help of Equation (1.84), a new type of fractional differential equations as shown in Equation (1.85) can be solved, where the boundary values for Equation (1.85) are set to be $y(0) = y'(0) = \cdots = y^{(N)}(0) = 0$ and $y(\infty) = y'(\infty) = \cdots = y^{(N)}(\infty) = 0$:

$$\sum_{l=0}^{N} a_l t^{\alpha+l} D^{\alpha+l} y(t) = u(t). \tag{1.85}$$

The Mellin transform method is considered as a potent substitution to the Laplace transform method when the fractional equation has a typical form as shown in Equation (1.85). Let us use an example to illustrate how to apply the Mellin transform method. Suppose a differential equation is given in Equation (1.86) and the boundary conditions are $y(0) = y'(0) = y^{(2)}(0) = 0$ and $y(\infty) = y'(\infty) = y^{(2)}(\infty) = 0$:

$$t^{\alpha+2} D^{\alpha+2} y(t) + t^{\alpha+1} D^{\alpha+} y(t) + t^{\alpha} D^{\alpha} y(t) = u(t). \tag{1.86}$$

Taking Mellin transform on both sides of Equation (1.86), we have

$$Y(z)\,\Gamma(1-z)\left[\frac{1}{\Gamma(1-z-\alpha)} + \frac{1}{\Gamma(-z-\alpha)} + \frac{1}{\Gamma(-z-\alpha-1)}\right] = U(z)$$

$$Y(z) = U(z)\,G(1-z) = U(z)\,\frac{\Gamma(1-z-\alpha)}{\Gamma(1-z)} \cdot \frac{1}{1+(z+\alpha)^2}$$

$$G(z) = \frac{\Gamma(z-\alpha)}{\Gamma(z)} \cdot \frac{1}{1+(1-z+\alpha)^2}.$$

Performing the inverse Mellin transform and, utilizing the residue theorem to invert the above transfer function, yields

$$g(t) = \frac{1}{2\pi j} \int_{\zeta-j\infty}^{\zeta+j\infty} G(z) t^{-z}\, dz$$

$$= \frac{1}{2\pi j} \cdot (2\pi j) \sum_{n=0}^{\infty} [Res\, G(z)t^{-z}]_{z=\alpha-n}.$$

From the complex analysis, we know the residue of $\Gamma(z)$ at ordinary poles $z = -n$, where n is an integer, is $Res\,[\Gamma(z)]_{z=-n} = \frac{(-1)^n}{n!}$. Therefore,

$$g(t) = \sum_{n=0}^{\infty} [Res\, G(z)\, t^{-z}]_{z=\alpha-n}$$

$$= \sum_{n=0}^{\infty} \left[Res\, \frac{\Gamma(z-\alpha)}{\Gamma(z)} \cdot \frac{t^{-z}}{1 + (1 - z + \alpha)^2} \right]_{z=\alpha-n}$$

$$= \sum_{n=0}^{\infty} \frac{(-1)^n\, t^{n-\alpha}}{n!\,\Gamma(\alpha - n)\left[1 + (n+1)^2 \right]}.$$

Applying the convolution theorem to the Mellin transform will give the final result as follows:

$$y(t) = \int_0^{\infty} u(t\tau)\, g(\tau)\, d\tau.$$

1.3.3.3 Fractional Green's Function

In the ordinary calculus, Green's function is assumed to be the impulse response with respect to an inhomogeneous linear differential equation, i.e., if we assume $_0^I\mathcal{L}_t$ to be a integer-order linear differential operator, then the Green's function satisfies Equation (1.87), where $\delta(t)$ is the Dirac function:

$$_0^I\mathcal{L}_t\,(G(t,\tau)) = \delta(t - \tau). \tag{1.87}$$

Consider a linear differential equation as shown in Equation (1.88), where $y(t)$ is the output and $u(t)$ is the input:

$$_0^I\mathcal{L}_t\,(y(t)) = u(t). \tag{1.88}$$

Based on Equations (1.87) and (1.88), the property of the Dirac function gives

$$_0^I\mathcal{L}_t\,(y(t)) = u(t) = \int \delta(t - \tau)\, u(\tau)\, d\tau = \int {}_0^I\mathcal{L}_t\,(G(t,\tau))\, u(\tau)\, d\tau$$

$$= {}_0^I\mathcal{L}_t\left(\int G(t,\tau)\, u(\tau)\, d\tau \right). \tag{1.89}$$

Therefore, the solution of a general linear differential equation can be expressed as a convolution integral of the Green's function and the input signal, i.e.,

$$y(t) = \int G(t,\tau)\, u(\tau)\, d\tau. \tag{1.90}$$

Green function is proposed for the solution of integer-order calculus. However, we can modify its definition so that it fits the cases of solving the linear fractional differential equations.

Let us consider a class of fractional differential equations described in Equation (1.91), which is the generalization of all types of linear fractional differential equations:

$$\substack{F\\0}\mathcal{L}_t y\,(t) = u(t) \tag{1.91}$$

$$\left[\substack{MR\\0}D_t^{\sigma_i-1} y\,(t)\right]_{t=0} = b_i,\ (i = 1,\dots,n),$$

where in Equation (1.91), the linear fractional differential operator $\substack{F\\0}\mathcal{L}_t$ is defined as the linear combination of sequential Miller–Ross fractional derivative operators.

$$\substack{F\\0}\mathcal{L}_t \stackrel{def}{=} \substack{MR\\0}D_t^{\sigma_n} + a_1\,(t)\,\substack{MR\\0}D_t^{\sigma_{n-1}} + \cdots + a_{n-1}\,(t)\,\substack{MR\\0}D_t^{\sigma_1} + a_n(t)$$

$$\substack{MR\\0}D_t^{\sigma_i} = \left(\substack{RL\\0}D_t^{\alpha_i}\right)\left(\substack{RL\\0}D_t^{\alpha_i-1}\right)\cdots\left(\substack{RL\\0}D_t^{\alpha_1}\right)$$

$$\substack{MR\\0}D_t^{\sigma_i-1} = \left(\substack{RL\\0}D_t^{\alpha_i-1}\right)\left(\substack{RL\\0}D_t^{\alpha_i-1}\right)\cdots\left(\substack{RL\\0}D_t^{\alpha_1}\right)$$

$$\sigma_i = \sum_{j=1}^{i}\alpha_j,\ (i = 1,\dots,n)\,;\, 0 \le \alpha_j \le 1,\ (j = 1,\dots,n).$$

To solve the linear fractional differential equation which has a form of Equation (1.91), we need to give the definition of the fractional Green's function $FG(t,\tau)$, which should satisfy the following two conditions [Cheng, 2013]:

$$\substack{F\\\tau}\mathcal{L}_t FG\,(t,\tau) = 0 \quad \text{for every} \quad \tau \in (0,t)$$

$$\lim_{\tau\to t-0}\left(\substack{MR\\\tau}D_t^{\sigma_i-1} FG(t,\tau)\right) = \delta_{i,n},\ i = 1,\dots,n.$$

Analogous to the integer-order Green's function, the fractional Green's function for Equation (1.91) with constant coefficients is $FG\,(t,\tau) = FG(t-\tau)$. And the solution to Equation (1.91) with $a_i\,(t) = a_i,\ i = 1,\dots,n$ is given below [Cheng, 2013]:

$$y\,(t) = \int_0^t FG\,(t-\tau)\,u\,(\tau)\,d\tau + \sum_{i=1}^{n} b_i\,\substack{MR\\0}D_t^{\sigma_n-\sigma_i} FG(t)$$

$$+ \sum_{i=1}^{n-1} a_i \sum_{j=1}^{n-i} b_j\,\substack{MR\\0}D_t^{\sigma_{n-i}-\sigma_j} FG(t). \tag{1.92}$$

Proof of Equation (1.92). In the previous section, we do not give the Laplace transform of the Miller–Ross fractional derivatives. However, we can get it by repeatedly applying the Laplace transform for the Riemann–Liouville fractional derivative in Equation (1.60), i.e.,

$$\mathcal{L}\left\{\substack{MR\\0}D_t^{\sigma_m} f(t)\right\} = s^{\sigma_m} F\,(s) - \sum_{i=0}^{m-1} s^{\sigma_m-\sigma_{m-i}}\left[\substack{MR\\0}D_t^{\sigma_m-i-1} f\,(t)\right]_{t=0}. \tag{1.93}$$

Taking the Laplace transform on both sides of Equation (1.91), we can get the following formula:

$$Y(s) = \frac{U(s)}{D(s)} + \frac{\sum_{i=1}^{n} b_i s^{\sigma_n - \sigma_i}}{D(s)} + \frac{\sum_{i=1}^{n-1} a_i \sum_{j=1}^{n-i} b_j s^{\sigma_n - i - \sigma_j}}{D(s)}, \qquad (1.94)$$

where $D(s) = s^{\sigma_n} + \sum_{i=1}^{n-1} a_i s^{\sigma_n - i} + a_n$. Then if we let $FG(t) = \mathcal{L}^{-1}\left\{\frac{1}{D(s)}\right\}$, the inverse Laplace transform of Equation (1.94) renders Equation (1.92). □

The reason why we believe that the inverse Laplace transform of the reciprocal of the denominator polynomial in the system transfer function is the fractional Green's function is proved as follows. It is clear that this asserted formula satisfies fractional Green's function's two conditions.

First, let us consider the Riemann–Liouville fractional derivative of the variable upper limit integral. The differentiation of an integral with variable upper limit is given in Equation (1.95):

$$\frac{d}{dt} \int_0^t f(t, \tau) \, d\tau = \int_0^t \frac{\partial f(t, \tau)}{\partial t} \, d\tau + \lim_{\tau \to t-0} f(t, \tau). \qquad (1.95)$$

A corollary formula in fractional case is given in Equation (1.96) [Podlubny, 1999]:

$$^{RL}_0 D_t^\alpha \int_0^t f(t, \tau) \, d\tau = \int_0^t {}^{RL}_\tau D_t^\alpha f(t, \tau) \, d\tau + \lim_{\tau \to t-0} {}^{RL}_\tau D_t^{\alpha-1} f(t, \tau). \qquad (1.96)$$

Proof. The proof to show that $\int_0^t FG(t - \tau) u(\tau) \, d\tau$ is the solution to Equation (1.91) with zero initial states is given as follows:

$$^{MR}_0 D_t^{\sigma_1} y(t) = {}^{RL}_0 D_t^{\alpha_1} \int_0^t FG(t - \tau) u(\tau) \, d\tau$$

(by Equation (1.96))

$$= \int_0^t {}^{RL}_\tau D_t^{\alpha_1} FG(t - \tau) u(\tau) \, d\tau + \lim_{\tau \to t-0} {}^{RL}_\tau D_t^{\alpha_1 - 1} FG(t - \tau) u(\tau)$$

(by Condition (b) of $FG(t)$)

$$= \int_0^t {}^{MR}_\tau D_t^{\sigma_1} FG(t - \tau) u(\tau) \, d\tau$$

$$^{MR}_{\ 0}D_t^{\sigma_2} y(t) = {}^{RL}_{\ 0}D_t^{\alpha_2} \int_0^t {}^{MR}_{\ \tau}D_t^{\sigma_1} FG(t-\tau) u(\tau)\, d\tau$$

$$= \int_0^t {}^{RL}_{\ \tau}D_t^{\alpha_2\, MR}_{\ \ \ \tau}D_t^{\sigma_1} FG(t-\tau) u(\tau)\, d\tau$$

$$+ \lim_{\tau \to t-0} {}^{RL}_{\ \tau}D_t^{\alpha_2-1\, MR}_{\ \ \ \tau}D_t^{\sigma_1} FG(t-\tau) u(\tau)$$

$$= \int_0^t {}^{MR}_{\ \tau}D_t^{\sigma_2} FG(t-\tau) u(\tau)\, d\tau$$

$$\vdots$$

$$\vdots$$

$$^{MR}_{\ 0}D_t^{\sigma_{n-1}} y(t) = {}^{RL}_{\ 0}D_t^{\alpha_{n-1}} \int_0^t {}^{MR}_{\ \tau}D_t^{\sigma_{n-2}} FG(t-\tau) u(\tau)\, d\tau$$

$$= \int_0^t {}^{RL}_{\ \tau}D_t^{\alpha_{n-1}\, MR}_{\ \ \ \tau}D_t^{\sigma_{n-2}} FG(t-\tau) u(\tau)\, d\tau$$

$$+ \lim_{\tau \to t-0} {}^{RL}_{\ \tau}D_t^{\alpha_{n-1}-1\, MR}_{\ \ \ \tau}D_t^{\sigma_{n-2}} FG(t-\tau) u(\tau)$$

$$= \int_0^t {}^{MR}_{\ \tau}D_t^{\sigma_{n-1}} FG(t-\tau) u(\tau)\, d\tau$$

$$^{MR}_{\ 0}D_t^{\sigma_n} y(t) = {}^{RL}_{\ 0}D_t^{\alpha_n} \int_0^t {}^{MR}_{\ \tau}D_t^{\sigma_{n-1}} FG(t-\tau) u(\tau)\, d\tau$$

$$= \int_0^t {}^{RL}_{\ \tau}D_t^{\alpha_n\, MR}_{\ \ \ \tau}D_t^{\sigma_{n-1}} FG(t-\tau) u(\tau)\, d\tau$$

$$+ \lim_{\tau \to t-0} {}^{RL}_{\ \tau}D_t^{\alpha_n-1\, MR}_{\ \ \ \tau}D_t^{\sigma_{n-1}} FG(t-\tau) u(\tau)$$

$$= \int_0^t {}^{MR}_{\ \tau}D_t^{\sigma_n} FG(t-\tau) u(\tau)\, d\tau + u(t).$$

Then, applying the linear fractional differential operator to $y(t)$, we have the following derivation which shows the result:

$$^F_0\mathcal{L}_t y(t) = {}^{MR}_{\ \ 0}D_t^{\sigma_n} y(t) + a_1 {}^{MR}_{\ \ 0}D_t^{\sigma_{n-1}} y(t) + \cdots + a_{n-1} {}^{MR}_{\ \ 0}D_t^{\sigma_1} y(t) + a_n y(t)$$

$$= u(t) + \int_0^t {}^{MR}_{\tau}D_t^{\sigma_n} FG(t-\tau)u(\tau)\,d\tau$$

$$+ a_1 \int_0^t {}^{MR}_{\tau}D_t^{\sigma_{n-1}} FG(t-\tau)u(\tau)\,d\tau + \cdots +$$

$$a_{n-1} \int_0^t {}^{MR}_{\tau}D_t^{\sigma_1} FG(t-\tau)u(\tau)\,d\tau + a_n \int_0^t FG(t-\tau)u(\tau)\,d\tau$$

$$= u(t) + \int_0^t {}^F_0\mathcal{L}_t FG(t-\tau)u(\tau)\,d\tau$$

(by Condition (b) of $FG(t)$)

$$= u(t).$$

Thus, $y(t) = \int_0^t FG(t-\tau)u(\tau)\,d\tau$ is indeed the solution to Equation (1.91) with zero initial states. And the assertion in the proof for Equation (1.92) is valid.   □

We already know that $\mathcal{L}^{-1}\left\{\frac{U(s)}{D(s)}\right\}$ is the solution to Equation (1.91) when the initial states $b_i, i = 1, \ldots, n$ are all zero. Thus, it can be further concluded that $FG(t) = \mathcal{L}^{-1}\left\{\frac{1}{D(s)}\right\}$ is true according to convolution theorem, which means our assertion in the proof for Equation (1.92) is valid.

1.3.4   NUMERICAL SOLUTIONS TO GENERAL-FORM FRACTIONAL DIFFERENTIAL EQUATIONS

For most natural fractional-order system plants which are governed by fractional differential equations, to which the analytical solutions may be difficult to be discovered or developed. In this subsection, we will present a brief introduction to the numerical approaches in solving the fractional differential equations. First, the numerical approximation for three definitions of the fractional derivative, i.e., Grünwald–Letnikov, Riemann–Liouville, and Caputo's fractional derivatives, is discussed. And in the second half part of this section, the numerical algorithms used to solve the general-form linear fractional differential equations with zero initial states are examined. For non-zero initial value (states) problems, we only consider the numerical methods to solve the linear Caputo's fractional differential equations. In this book, we do not study the computational algorithms to solve the non-zeros initial value problems which are modelled with the Riemann–Liouville or Grünwald–Letnikov fractional differential equations. This is because the three definitions for the fractional derivatives can be related to each other using an explicit formula, as we discussed in Sections 1.2.1 and 1.2.2. And the three definitions for the fractional derivatives are identical in the zero initial value problems, which can be seen from the equivalence between Equations (1.39) and (1.40).

1.3.4.1 High-Precision Approximation of Fractional-Order Derivatives

Riemann–Liouville fractional derivative and Grünwald–Letnikov fractional derivative are essentially equivalent if we assume function $f(t)$ is $(m+1)$-th order continuously differentiable. Therefore, we will only numerically compute the Grünwald–Letnikov fractional derivative, which will also be identical to the related Riemann–Liouville fractional derivative. By the definition of the Grünwald–Letnikov fractional derivative in Equation (1.23), the coefficients c_i in the series are given below:

$$c_i = (-1)^i \binom{\alpha}{i} = (-1)^i \frac{\Gamma(\alpha+1)}{\Gamma(i+1)\,\Gamma(\alpha-i+1)}. \tag{1.97}$$

And a recursive relation for the coefficient sequence is given as follows:

$$\frac{c_i}{c_{i-1}} = -\frac{\Gamma(i)\,\Gamma(\alpha-i+2)}{\Gamma(i+1)\,\Gamma(\alpha-i+1)} = 1 - \frac{\alpha+1}{i} \quad c_0 = 1. \tag{1.98}$$

A first-order approximation algorithm for the Grünwald–Letnikov fractional derivative and Riemann–Liouville fractional derivative is given below. In the algorithm block, $t_1 = 0$ is the initial time and y_i, $i = 1, \ldots, N$ is the discrete sequence of the Grünwald–Letnikov fractional derivative.

Algorithm 1.1 (First-order Grünwald–Letnikov or Riemann–Liouville fractional derivative and integral)

1: **procedure** :
2:      Choose proper parameters: the fixed time step size h, the total number of discrete time points N.
3:      For $p = 2, \ldots, N$, set

$$t_p = t_{p-1} + h$$

4:      For $i = 2, \ldots, N$, use the recursive relation in Equation (1.98) to obtain coefficients, set

$$c_i = c_{i-1}\left(1 - \frac{\alpha+1}{i-1}\right)$$

5:      For $j = 1, \ldots, N$, sum all the terms in Equation (1.23) to get the approximated Grünwald–Letnikov fractional derivative or integral,

$$y_i = \sum_{k=1}^{j} \frac{c_k\, f(t_{j-k+1})}{h^\alpha}$$

Since the Algorithm 1.1 is noted to have a first-order accuracy [Podlubny, 1999], using this method may lead to large deviation from the true value of the Grünwald–Letnikov fractional derivative or integral. To overcome its deficiency in accuracy, we present a general method which can generate algorithms with any order of accuracy. Let us first introduce a type of special functions as shown in Equation (1.99):

$$h_r(x) = \sum_{k=1}^{r} \frac{1}{k}(1-x)^k .$$

$$(1.99)$$

When we set $r = 1$, we can see that $(h_1(x))^\alpha = (1-x)^\alpha$ and the coefficients of its expansion series coincides with the coefficients c_i in the Algorithm 1.1. Correspondingly, if we require Algorithm 1.1 to have an accuracy order of r, we should modify the formula of coefficients c_i and make them coincide with the coefficients of $(h_r(x))^r$ expansion. This result has been proven by Lubich [1986]. Now, let us directly outline the improved algorithm as given in Algorithm 1.2.

Algorithm 1.2 (r^{th}-order Grünwald–Letnikov or Riemann–Liouville fractional derivative and integral)

1: **procedure** :
2:     Choose proper parameters: the fixed time step size h, the total number of discrete time points N.
3:     For $p = 2,\ldots,N$ with $t_1 = 0$, set

$$t_p = t_{p-1} + h$$

4:     For $i = 1,\ldots,N$, expand $(h_r(x))^\alpha = \left(\sum_{k=1}^{r} \frac{1}{k}(1-x)^k\right)^\alpha$ and suppose the coefficients of the expansion are d_i, $i = 1, 2, \ldots,$ set

$$c_i = d_i$$

5:     For $j = 1,\ldots,N$, sum all the terms in Equation (1.23) to get the approximated Grünwald–Letnikov fractional derivative or integral,

$$y_i = \sum_{k=1}^{j} \frac{c_k f(t_{j-k+1})}{h^\alpha}$$

The previous two algorithms are aimed at approximating both Grünwald–Letnikov and Riemann–Liouville fractional derivatives. Subsequently, we need to consider how to approximate the Caputo's fractional derivative or integral. As we discussed before, we rewrite the relationship between the Caputo's fractional derivative and the Riemann–Liouville fractional derivative in Equation (1.100), where $m \leq \alpha < m + 1$:

$$_{a}^{RL}D_t^{\alpha} f(t) = \sum_{k=0}^{m} \frac{f^{(k)}(a)(t-a)^{-\alpha+k}}{\Gamma(-\alpha+k+1)} + {}_{a}^{C}D_t^{\alpha} f(t) = {}_{a}^{GL}D_t^{\alpha} f(t). \tag{1.100}$$

Let $df_0 = \left[f(a), f'(a), \ldots, f^{(m)}(a) \right]$ be the initial values of the objective function $f(t)$, we can propose a high-precision algorithm to approximate the Caputo's fractional derivative or integral. The algorithm is outlined and given in Algorithm 1.3.

Algorithm 1.3 (Caputo's fractional derivative and integral)

1: **procedure** :
2:     Choose a proper one from Alogrithm 1.1 and Algorithm 1.2 to approximate the Grünwald–Letnikov fractional derivative or integral, say gly is the approximated value.
3:     For $i = 1, \ldots, N$, remove the truncation error term in Equation (1.100), the approximated Caputo's derivative or integral is given as

$$y_i = gly_i - \sum_{k=0}^{m} \frac{f^{(k)}(a)(t_i - a)^{-\alpha+k}}{\Gamma(-\alpha+k+1)}$$

To test the effectiveness of these three algorithms, let us consider an example with the objective function $f(t) = 10u(t)$, where $u(t)$ is a unit-step function, let us compute the Riemann–Liouville and Caputo's fractional derivative of it with fractional orders equal to 0.5 and -0.5. The Riemann–Liouville fractional derivative of $f(t)$ has an analytical formula which is given in Equation (1.101) and the Caputo's fractional derivative to a constant is zero:

$$_{0}^{RL}D_t^{\alpha}(10u(t)) = \frac{10t^{-\alpha}}{\Gamma(1-\alpha)}. \tag{1.101}$$

The algorithms are implemented in MATLAB. The simulation results are plotted in two figures below. Figure 1.4 shows the numerical 0.5th Grünwald–Letnikov fractional derivatives with first-order and second-order algorithms, compared with the exact derivative shown in Equation (1.101). Figure 1.5 presents the relevant simulation curves for 0.5-folded integral.

1.3.4.2 Numerical Solutions to a Fractional Differential Equations

- Zero state response of the fractional LTI system.

Let us consider the initial value problems which are governed by linear fractional differential equations with constant coefficients. The initial value problems are formulated in Equation (1.102), where $u(t)$ is the input signal and $y(t)$ is the output signal, their derivatives at

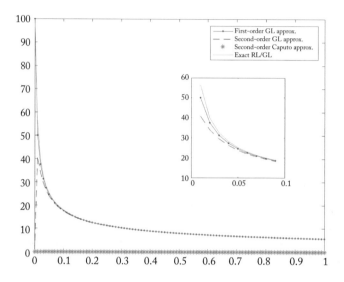

Figure 1.4: Numerical 0.5th fractional derivatives for a constant value function using different algorithms.

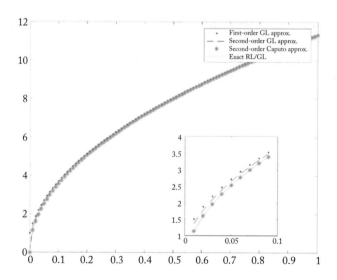

Figure 1.5: Numerical 0.5-folded fractional integral for a constant value function using different algorithms.

$t = 0$ are all constrained to be zero. Therefore, all different definitions for fractional operators are equivalent in Equation (1.102). For simplicity, let us denote D^α as the α-th order fractional derivative.

$$a_1 D^{\alpha_1} y(t) + a_2 D^{\alpha_2} y(t) + \cdots + a_n D^{\alpha_n} y(t) = b_1 D^{\beta_1} u(t) + $$
$$b_2 D^{\beta_2} u(t) + \cdots + b_m D^{\beta_m} u(t). \tag{1.102}$$

Since coefficients for the Grünwald–Letnikov derivative are given in Equation (1.97), the α_i-th order fractional operator operating on function $y(t)$ gives:

$$D^{\alpha_i} y(t) = \frac{1}{h^{\alpha_i}} \sum_{j=0}^{N} c_j^{(\alpha_i)} y(t - jh). \tag{1.103}$$

Assume $N = [t_f / h]$, t_f is the length of the simulation interval, and the sampled input, output signals are given as $u = [u_0, u_1, \ldots, u_N]^T$ and $y = [y_0, y_1, \ldots, y_N]^T$. Discretizing the input signal and output signal in the simulation time interval and expressing the post-discretized coefficients in matrix form will give Equation (1.106), which is the approximated difference equation for Equation (1.102) [Xue and Bai, 2016]:

$$A^{(\alpha_i)} = \frac{1}{h^{\alpha_i}} \begin{bmatrix} c_0^{(\alpha_i)} & 0 & 0 & 0 & \cdots & 0 \\ c_1^{(\alpha_i)} & c_0^{(\alpha_i)} & 0 & 0 & \cdots & 0 \\ c_2^{(\alpha_i)} & c_1^{(\alpha_i)} & c_0^{(\alpha_i)} & 0 & \cdots & 0 \\ \vdots & \vdots & \ddots & \ddots & \ddots & \vdots \\ c_{N-1}^{(\alpha_i)} & \cdots & c_2^{(\alpha_i)} & c_1^{(\alpha_i)} & c_0^{(\alpha_i)} & 0 \\ c_N^{(\alpha_i)} & c_{N-1}^{(\alpha_i)} & \cdots & c_2^{(\alpha_i)} & c_1^{(\alpha_i)} & c_0^{(\alpha_i)} \end{bmatrix}, \tag{1.104}$$

$$B^{(\beta_i)} = \frac{1}{h^{\beta_i}} \begin{bmatrix} c_0^{(\beta_i)} & 0 & 0 & 0 & \cdots & 0 \\ c_1^{(\beta_i)} & c_0^{(\beta_i)} & 0 & 0 & \cdots & 0 \\ c_2^{(\beta_i)} & c_1^{(\beta_i)} & c_0^{(\beta_i)} & 0 & \cdots & 0 \\ \vdots & \vdots & \ddots & \ddots & \ddots & \vdots \\ c_{N-1}^{(\beta_i)} & \cdots & c_2^{(\beta_i)} & c_1^{(\beta_i)} & c_0^{(\beta_i)} & 0 \\ c_N^{(\beta_i)} & c_{N-1}^{(\beta_i)} & \cdots & c_2^{(\beta_i)} & c_1^{(\beta_i)} & c_0^{(\beta_i)} \end{bmatrix}. \tag{1.105}$$

The discretized algebraic equation is given as:

$$\sum_{i=1}^{n} A^{(\alpha_i)} y = \sum_{j=1}^{m} B^{(\beta_j)} u. \tag{1.106}$$

Let us denote $A = \sum_{i=1}^{n} A^{(\alpha_i)}$, and $B = \sum_{j=1}^{m} B^{(\beta_j)}$, Equation (1.106) can be solved using numerical linear algebra techniques, such as iterative or direct inversion method. The most straightforward way is the least square method. In this case, the numerical solution to Equation (1.102) is given as $y = A^{-1}Bu$, where A^{-1} is pseudo-inverse of matrix A.

As per our discussion, the first-order finite difference method is used to discretize the fractional differential equation, which is as same as what we present in Algorithm 1.1. Since this way of discretization has low precision order, the truncation error between Equations (1.102) and (1.106) may not be ignorable. Thus, Algorithm 1.2 can be a much better choice to approximate the fractional operators in Equation (1.102). For a fractional linear time variant (LTV) system, the coefficients in Equation (1.102) will no longer be constant, and the nonconstant coefficients are also required to be discretized. The discretized LTV system equations can be solved using iterative methods. In the next part, the focus will be switched to the linear Caputo's fractional differential equations with nonzero initial conditions.

• Linear Caputo's fractional differential equations, with nonzero initial states.

Consider the linear Caputo's fractional differential equation in Equation (1.107). Assume $\alpha_1 \geq \alpha_2 \geq \cdots \geq \alpha_n$ and $\beta_1 \geq \beta_2 \geq \cdots \geq \beta_m$. The initial states are $y(0) = y_0, y'(0) = y_0', \ldots, y^{(p)}(0) = y_0^{(p)}$, where p is the largest integer which is less or equal to α_1:

$$a_1 {}_0^C D_t^{\alpha_1} y(t) + a_2 {}_0^C D_t^{\alpha_2} y(t) + \cdots + a_n {}_0^C D_t^{\alpha_n} y(t)$$
$$= b_1 {}_0^C D_t^{\beta_1} u(t) + b_2 {}_0^C D_t^{\beta_2} u(t) + \cdots + b_m {}_0^C D_t^{\beta_m} u(t). \tag{1.107}$$

In order to solve Equation (1.107) with nonzero initial states, a compensation function $C(t)$ can be introduced to convert the original problem into a linear fractional differential equation with zero initial states:

$$C(t) = \sum_{i=0}^{p} \frac{y^{(i)}(0)}{i!} t^i. \tag{1.108}$$

Substituting $y(t) = x(t) + C(t)$ into Equation (1.107) yields a linear fractional differential equation with respect to $x(t)$. It can be verified that the initial values for this new equation are all zeros. The equivalent zero initial states problem to Equation (1.107) is given in Equation (1.109):

$$a_1 {}_0^C D_t^{\alpha_1} x(t) + a_2 {}_0^C D_t^{\alpha_2} x(t) + \cdots + a_n {}_0^C D_t^{\alpha_n} x(t)$$
$$= b_1 {}_0^C D_t^{\beta_1} u(t) + b_2 {}_0^C D_t^{\beta_2} u(t) + \cdots + b_m {}_0^C D_t^{\beta_m} u(t) - M(t), \tag{1.109}$$

where $M(t) = a_1 {}_0^C D_t^{\alpha_1} C(t) + a_2 {}_0^C D_t^{\alpha_2} C(t) + \cdots + a_n {}_0^C D_t^{\alpha_n} C(t)$.

• Nonlinear Caputo's fractional differential equations.

Let us consider a special class of nonlinear Caputo's fractional differential equations, as described in Equation (1.110). We emphasize to solve this class of nonlinear fractional equations because the existence and uniqueness theorem for its solutions has already been set up in the previous discussions:

$$_0^C D_t^\alpha y(t) = f(t, y(t)), \qquad (1.110)$$

where the accentuated $y(t)$ is the vectorized variable and $f(t, y(t))$ is the nonlinear vectorized function. The initial value for this equation is $y(0) = y_0$. p is the largest integer which is less or equal to α.

In order to solve this type of nonlinear fractional differential equation, we apply the fractional Adam–Bashforth algorithm. The algorithm is developed as follows.

We first use Laplace transform method to get the analytical solution to this equation:

$$\mathcal{L}\left\{_0^C D_t^\alpha y(t)\right\} = s^\alpha Y(s) - \sum_{k=0}^{p-1} s^{\alpha-k-1} y^{(k)}(0) = \mathcal{L}\{f(t, y(t))\}.$$

Therefore, the Laplace transform of $y(t)$ is

$$Y(s) = \sum_{k=0}^{p-1} s^{-k-1} y^{(k)}(0) + \frac{1}{s^\alpha} \mathcal{L}\{f(t, y(t))\}.$$

Taking the inverse Laplace transform on both sides of the equation above, we can get the analytical solution to Equation (1.110):

$$y(t) = \sum_{k=0}^{p-1} y^{(k)}(0)\frac{t^k}{k!} + \frac{1}{\Gamma(\alpha)} \int_0^t \frac{f(\tau, y(\tau))}{(t-\tau)^{1-\alpha}} d\tau. \qquad (1.111)$$

Discretize the output signal in simulation interval with step size h, that is, $t_j = jh$ and $y(t_j) = y_j$, the fractional Adam–Bashforth Algorithm is outlined and presented as Algorithm 1.4.

Algorithm 1.4 (Fractional Adam–Bashforth Algorithm)

1: **procedure** :

2:   For $j = 0, 1, \ldots$, compute the coefficients for predictor formula:

$$s_{i,j+1} = \frac{h^\alpha}{\alpha} \left[(j + 1 - i)^\alpha - (j - i)^\alpha \right]$$

$$\breve{y}_{j+1} = \sum_{i=0}^{p-1} y_i \frac{t_{j+1}^i}{i!} + \frac{1}{\Gamma(\alpha)} \sum_{i=0}^{j} s_{i,j+1} f(t_i, y_i)$$

3:   Compute the coefficients and update the $(k + 1)$-th step:

$$d_{0,j+1} = \frac{h^\alpha \left[j^{\alpha+1} - (j - \alpha)(j + 1)^\alpha \right]}{\alpha(\alpha + 1)},$$

$$d_{i,j+1} = \frac{h^\alpha \left[(j - i + 2)^{\alpha+1} + (j - i)^{\alpha+1} - 2(j - i + 1)^{\alpha+1} \right]}{\alpha(\alpha + 1)},$$

$$d_{j+1,j+1} = \frac{h^\alpha}{\alpha(\alpha + 1)},$$

$$y_{j+1} = \sum_{i=0}^{p-1} y_i \frac{t_{j+1}^i}{i!} + \frac{1}{\Gamma(\alpha)} \left[d_{j+1,j+1} f\left(t_{j+1}, \breve{y}_{j+1}\right) + \sum_{i=0}^{j} d_{i,j+1} f(t_i, y_i) \right],$$

4:   Test whether the convergence criterion of norm $\| y_{j+1} - \breve{y}_{j+1} \|$ is satisfied:

5:   **if** $\| y_{j+1} - \breve{y}_{j+1} \| < \epsilon$, **then** stop the iteration;
     Else, go to step 2.

C H A P T E R 2

Fractional-Order Controller Design

2.1 FREQUENCY DOMAIN DESIGN TECHNIQUES

As we have learned in integer-order linear control theory, the transfer function is a useful way to describe linear system plants. Bode plots can deliver the frequency response of the derivative or integral operator, with the magnitude of the transfer function versus the frequency sloping downward with slope $20n$ dB/dec and phase lead of $n\pi/2$ rad. Similarly, the fractional-order controller can also exhibit a wealthy information in frequency domain and its research has drawn more and more attention recently. In this chapter, a formal definition for the fractional-order transfer function will be presented. Consider the simplest form of a fractional-order linear operator, D^α. No matter what type of fractional derivative definition it adopts, the transfer function will be the same if the zero initial condition is preconditioned. Combined by these fractional-order linear operators, the linear fractional-order transfer function can be formulated by employing the Laplace transform techniques. In the second section, we will introduce Oustaloup filter to approximate the fractional linear operators and linear fractional-order transfer functions [Oustaloup, 1995]. Finally, the conventional PID controller will be generalized to its fractional order.

2.1.1 FRACTIONAL-ORDER TRANSFER FUNCTION

Equation (1.63) gives the fractional-order transfer function for a fractional-order LTI system. If a time delay T is considered, the linear fractional-order transfer function is given in Equation (2.1). Similarly, a fractional-order commensurate system has its general-form transfer function shown in Equation (2.2) [Xue, 2017]:

$$G\left(s\right) = \frac{Y\left(s\right)}{U\left(s\right)} = \frac{b_1 s^{\beta_1} + b_2 s^{\beta_2} + \cdots + b_m s^{\beta_m}}{a_1 s^{\alpha_1} + a_2 s^{\alpha_2} + \cdots + a_n s^{\alpha_n}} e^{-Ts} \tag{2.1}$$

$$G\left(s\right) = \frac{b_1 s^{m\alpha} + b_2 s^{(m-1)\alpha} + \cdots + b_m s^{\alpha} + b_{m+1}}{a_1 s^{n\alpha} + a_2 s^{(n-1)\alpha} + \cdots + a_n s^{\alpha} + a_{n+1}}. \tag{2.2}$$

For a multi-input multi-output (MIMO) system, the system transfer function will be given in a matrix form if we assume there are M inputs and N ourtputs. In Equation (2.3),

$G_{ij}(s)$ is the sub-transfer function from the jth input to the ith output signal:

$$G(s) = \begin{bmatrix} G_{11}(s) & \cdots & G_{1M}(s) \\ \vdots & \ddots & \vdots \\ G_{N1}(s) & \cdots & G_{NM}(s) \end{bmatrix}. \tag{2.3}$$

In this book, we will concentrate our attention on the analysis for the single-input single-output (SISO) system, in that a SISO transfer function is the fundamental component of the multiple-input multiple-output (MIMO) system transfer function matrix. For example, if we can assert that $G_{ij}(s)$ is stable for $i = 1, \ldots, N$ and $j = 1, \ldots, M$, then we can promise the overall stability of the MIMO system in Equation (2.3).

2.1.1.1 Open-Loop Stability Criterion in Frequency Domain

Neglecting the time-delay effects, we can analyze the stability of the linear commensurate system in regards to its transfer function in Equation (2.2), which can be written in the following form if $\zeta = s^\alpha$ is substituted into Equation (2.2):

$$G(\zeta) = \frac{b_1 \zeta^m + b_2 \zeta^{m-1} + \cdots + b_m \zeta + b_{m+1}}{a_1 \zeta^n + a_2 \zeta^{n-1} + \cdots + a_n \zeta + a_{n+1}}. \tag{2.4}$$

The commensurate linear fractional system is stable if we constrain the argument angle of the poles $\frac{\pi}{2} < |\arg(s)| < \pi$ in s domain or $\frac{\alpha\pi}{2} < |\arg(\zeta)| < \alpha\pi$. Stable regions and instable regions are depicted in Figure 2.1.

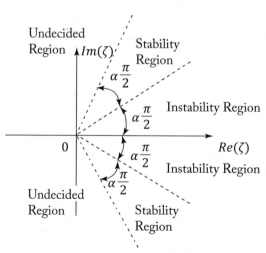

Figure 2.1: Stability region for a commensurate linear fractional system.

By observing the commensurate linear fractional transfer function shown in Equation (2.4), we can extract the denominator polynomial as the characteristic polynomial of the system. The zeros of the characteristic polynomial are the poles of the commensurate fractional system. That is, we can determine whether the system is stable by solving the zeros of the characteristic polynomial of the system.

The stability for a general linear fractional system can be tested via its commensurate system in form of Equation (2.2). First, we need to convert the general fractional transfer function in Equation (2.2) into Equation (2.4), with the base order valued as the greatest common divisor (gcd) of all the orders in the denominator and numerator polynomials, i.e., $\alpha = gcd(\alpha_1, \ldots, \alpha_n, \beta_1, \ldots, \beta_m)$. Note that α should be taken as a devisor between $0 < \alpha < 1$. Second, substitute $\zeta = s^\alpha$ into the commensurate-form transfer function to get the ζ-form transfer function. Eventually, solve the zeros of the denominator polynomials and assess the stability of the system based on the stability criterion.

Let us illustrate this treatment with examples as follows.

Consider a linear fractional system with its transfer function in Equation (2.5):

$$G(s) = \frac{s^{1.8} + 6s^{1.2} + 10s^{0.6} + 12}{2.5s^{2.4} + 5s^{1.8} + 4s^{1.2} + 7.5s^{0.6} + 12}. \tag{2.5}$$

The characteristic polynomial in ζ domain is given as Equation (2.6) with base order $\alpha = 0.6$. Solve the zeros of this polynomial and we can obtain the genuine poles to the original system:

$$p(\zeta) = 2.5\zeta^4 + 5\zeta^3 + 4\zeta^2 + 7.5\zeta + 12. \tag{2.6}$$

Two of the four poles of this open-loop system transfer function are portrayed in the ζ domain as two red asterisks shown in Figure 2.2. As we can see, the system with transfer function in Equation (2.5) possess two genuine poles located within the stability region, meaning the open-loop system is stable according to our stability criterion.

Even though the given stability criterion provides us with a direct passage to assess the stability of a linear fractional system, there still exist some limitations. For example, if the linear fractional system does not hold any genuine poles or instable poles, in other words, if all the poles of the system are lying within the undecided region, we are not able to tell the stability status of the system. Let us see the example in Equation (1.75) which we have solved in the previous chapter; it is obvious that the transfer function of this system is stable by observing the plot for the analytical output in Figure 1.3. However, if we evaluate the stability using the criterion introduced in this subsection, we find the linear fractional system of Equation (1.75) have poles at $\zeta = -1, -2, -3$, and -4, which are all lying in the undecided region. It means the stability may still be conditionally guaranteed if the system holds no genuine poles within the stability region but holds extraneous poles within the undecided region, and the system is exposed to small amount of disturbances. In the next section, we will reexamine the stability of the linear fractional system from the viewpoint of the state space.

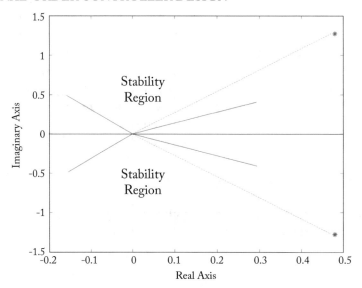

Figure 2.2: Open-loop pole map for the example in Equation (2.5).

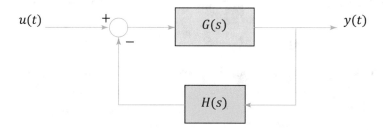

Figure 2.3: Black diagram of the negative unit feedback control.

2.1.1.2 Closed-Loop Stability Criterion, Nyquist Plot, and Bode Plot

Consider a negative unit feedback control, the forward system plant is $G(s)$, the feedback plant is $H(s) = 1$, the closed-loop system block diagram is sketched in Figure 2.3, and the closed-loop transfer function $G_c(s)$ is assessed in Equation (2.7). Based on the closed-loop transfer function, we are able to test the stability for the closed-loop system using the stability criterion as same as that for the open-loop transfer function. Sometimes, the expression for the closed loop transfer function is too complicated to be computed and we are introducing a new criterion to test the stability for the closed-loop system, that is, the Nyquist stability criterion for the linear fractional systems:

$$G_c(\zeta) = \frac{G(\zeta)}{1 + G(\zeta)}. \tag{2.7}$$

The Nyquist contour is encircling the area $-\frac{\alpha\pi}{2} < \arg(\zeta) < \frac{\alpha\pi}{2}$ in clockwise within Figure 2.1, it is mapped through the function $1 + G(\zeta)$ to the complex image plane and form a contour Γ_{1+G}, and the Nyquist contour is mapped though the function $G(\zeta)$ to the complex image plane and form a contour Γ_G. Let Z denote the number of zeros of $1 + G(\zeta)$ and P denote the number of poles of $1 + G(\zeta)$; then the contour Γ_{1+G} will encircle the origin point $N = Z - P$ times in clockwise and the contour Γ_G will encircle point $-1 + j0$ for N times in clockwise as well. It should be noted that Z is also the number of poles of the closed-loop transfer function $G_c(\zeta)$ and P is also the number of poles of the open-loop transfer function $G(\zeta)$.

The Nyquist stability criterion tells us: if the open loop system has P unstable poles, the Nyquist plot Γ_G must encircle point $-1 + 0j$ for P times in counterclockwise in order to make the closed-loop system stable. As follows, let us test the closed-loop stability of the system transfer function in Equation (2.5). The open loop transfer function is stable, which mean $P = 0$. The Nyquist plot Γ_G is sketched in Figure 2.4, which shows the contours do not encircle the point $-1 + j0$ in clockwise. Therefore, the number of unstable poles for the closed-loop transfer function is $Z = 0$, and this guarantees the stability of the closed-loop system.

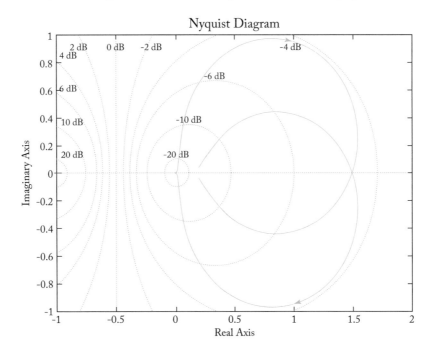

Figure 2.4: Nyquist plot of the example in Equation (2.5).

To check the correctness of the results derived from the Nyquist stability criterion, let us formulate the closed-loop transfer function for the system in Equation (2.5), which is given in

Equation (2.8). The poles of the closed-loop transfer function is plotted in Figure 2.5. It shows that the closed-loop transfer function has two stable poles. Therefore, the results derived from the Nyquist stability criterion are fully right:

$$G\,(s) = \frac{s^{1.8} + 6s^{1.2} + 10s^{0.6} + 12}{2.5s^{2.4} + 6s^{1.8} + 10s^{1.2} + 17.5s^{0.6} + 24}. \tag{2.8}$$

Besides the Nyquist plot, the Bode plot also plays an important role in the frequency analysis for the linear fractional system. By observing Equation (2.7), if $|G(j\omega)| = 1$ and $\angle G\,(j\omega) = -180°$, the closed-loop system will be unstable. The Bode plot for the example transfer function (Equation (2.5)) is shown in Figure 2.6. The phase margin is the difference of phase at the gain crossover frequency to the $-180°$, and the gain margin is the difference of 0 dB to the gain at the phase crossover frequency. The gain crossover frequency is 2.6405 rad/s and the phase margin is 99.7802° in Figure 2.6. Therefore, the frequency analysis with the Bode plot gives the same stability results as the results given by the Nyquist stability analysis for the example transfer function (Equation (2.5)).

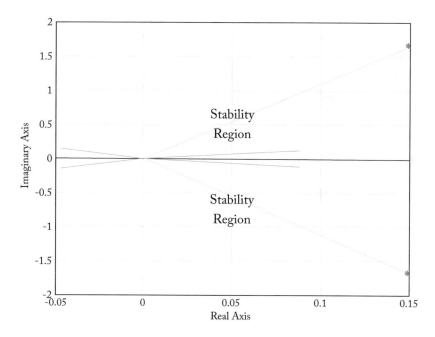

Figure 2.5: Closed-loop pole map for the example in Equation (2.5).

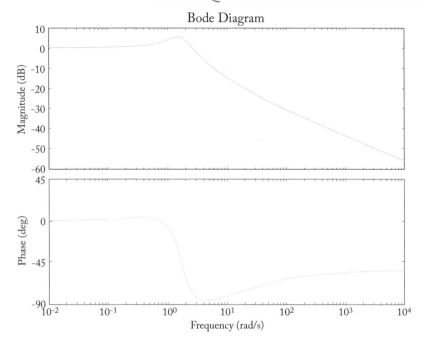

Figure 2.6: Bode plot for the example in Equation (2.5).

2.1.2 APPROXIMATION OF FRACTIONAL-ORDER TRANSFER FUNCTION

With the frequency domain analysis, the properties and performance behaviors of a linear fractional system plant can be predicted by its fractional transfer function. Given a fractional transfer function, it is quite straightforward to draw its Nyquist plot and Bode plot by setting $s = j\omega$ in the transfer function. However, an integer-order approximation for a linear fractional plant needs to be employed in the Simulink simulations, which is due to the lack of fractional-derivative functions defined in the MATLAB kernel. In this section, we will introduce the Oustaloup filter to approximate the linear fractional operators and linear fractional transfer functions in frequency domain. As a result, the approximated Oustaloup filter can be inserted into the Simulink model to function as a linear fractional system plant. The Oustaloup filter was proposed by French mathematician Alain Oustaloup, whose basic idea was to use a zig-zag gain curve to fit the fractional gain curve in a Bode plot. This idea of Oustaloup fitting is plotted in Figure 2.7. In the figure, blue line represents the gain curve for the fractional operator s^{α}, which has a slope equal to -20α dB/dec. The red zig-zag curve is the Oustaloup filter curve to approximate a fractional operator, and its general form can be expressed in Equation (2.9):

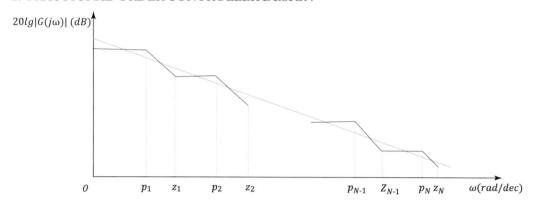

Figure 2.7: Oustaloup fitting in a Bode gain plot.

$$s^\alpha \approx K \prod_{i=1}^{N} \frac{s + z_i}{s + p_i}. \tag{2.9}$$

Suppose the interest frequency range is $\omega \in (\omega_l, \omega_u)$; the parameters in Equation (2.9) are determined by following formulas:

$$z_i = \omega_l \omega_m^{\frac{2i-1-\alpha}{N}}, \quad p_i = \omega_l \omega_m^{\frac{2i-1+\alpha}{N}}, \quad K = \omega_u^\alpha,$$

$$\omega_m = \sqrt{\frac{\omega_u}{\omega_l}}, \quad \text{for} \quad i = 1, 2, \ldots, N. \tag{2.10}$$

The original Oustaloup filter shows great approximation between the interested frequency range, but the approximation at the outliers are very inaccurate. A modified Oustaloup filter was proposed to improve the outlier fitting. The filter has a form in Equation (2.11), where $a = 10$ and $b = 9$ are default values for the two inherent parameters a and b. Parameters ω_u, z_i, and p_i are also given in Equation (2.10):

$$s^\alpha \approx \left(\frac{b\omega_u}{a}\right)^\alpha \left(\frac{bs^2 + a\omega_u s}{b(1 - \alpha)s^2 + a\omega_u s + b\alpha}\right) \prod_{i=1}^{N} \frac{s + z_i}{s + p_i}. \tag{2.11}$$

Note that the fractional order α should be limited in $(-1, 1)$, set $N = 5, \omega_l = 0.01$ and $\omega_u = 1000$. Let us use Equation (2.9) to approximate the linear fractional transfer function in

Equation (2.5):

$$s^{0.6} \approx \frac{63.1s^5 + 1.111 \times 10^4 s^4 + 1.779 \times 10^5 s^3 + 2.819 \times 10^5 s^2 + 4.423 \times 10^4 s + 631}{s^5 + 701.1s^4 + 4.468 \times 10^4 s^3 + 2.819 \times 10^5 s^2 + 1.761 \times 10^5 s + 10^4}$$

$$s^{1.2} \approx \frac{3.981s^6 + 1111s^5 + 2.819 \times 10^4 s^4 + 7.081 \times 10^4 s^3 + 1.761 \times 10^4 s^2 + 398.1s}{s^5 + 442.3s^4 + 1.779 \times 10^4 s^3 + 7.081 \times 10^4 s^2 + 2.791 \times 10^4 s + 1000}$$

$$s^{1.8} \approx \frac{251.2s^6 + 3.514 \times 10^4 s^5 + 4.468 \times 10^5 s^4 + 5.624 \times 10^5 s^3 + 7.011 \times 10^4 s^2 + 794.3s}{s^5 + 882.6s^4 + 7.081 \times 10^4 s^3 + 5.624 \times 10^5 s^2 + 4.423 \times 10^5 s + 3.162 \times 10^4}$$

$$s^{2.4} \approx \frac{15.85s^7 + 3514s^6 + 7.081 \times 10^4 s^5 + 1.413 \times 10^5 s^4 + 2.791 \times 10^4 s^3 + 501.2s^2}{s^5 + 556.9s^4 + 2.819 \times 10^4 s^3 + 1.413 \times 10^5 s^2 + 7.011 \times 10^4 s + 3162}.$$

Substitute the approximated fractional derivative operators into Equation (2.5); the Oustaloup approximation for the linear fractional transfer function can be obtained. Similarly, we can also approximate Equation (2.5) using the modified Oustaloup filter shown in Equation (2.11).

The Bode plot for Equation (2.5), its Oustaloup approximation, and its modified Oustaloup approximation are given to see the performance of the approximation; see Figure 2.8. The filters are taking order to be $N = 5$, and the interested frequency range $\omega_l = 10^{-2}$ rad/s and $\omega_u = 10^3$ rad/s. The modified Oustaloup filter exhibits a more accurate approximation performance.

2.1.3 FRACTIONAL-ORDER PID CONTROLLERS

Proportional-integral-derivative controller, abbreviated as PID controller, is commonly used in industrial control. Usually, it has parameters K_p, K_i, and K_d to be set, along with an additional parameter $T_c = 0.001$ to promise the causality of the derivative term in the controller. The general form of the integer-order PID controller is given in Equation (2.12):

$$C(s) = K_p + \frac{K_i}{s} + \frac{K_d s}{T_c s + 1}. \tag{2.12}$$

The fractional-order PID controller, or $PI^\phi D^\psi$ controller, has a general form in Equation (2.13):

$$C_F(s) = K_p + \frac{K_i}{s^\phi} + K_d s^\psi. \tag{2.13}$$

Compared to the conventional PID controller $C(s)$, two more parameters ϕ and ψ need to be determined in the fractional-order PID controller $C_F(s)$. With more parameters to be tuned, the fractional-order PID controller may deliver a better control performance than PID controller.

For a typical negative unit feedback control system, as shown in Figure 2.9, we have $H(s) = 1$ and loop transfer function $L(s) = C_F(s) P(s) H(s) = C_F(s) P(s)$, the output disturbance $d(t)$, and the sensor noise $n(t)$ are also considered in the system. Therefore, we can

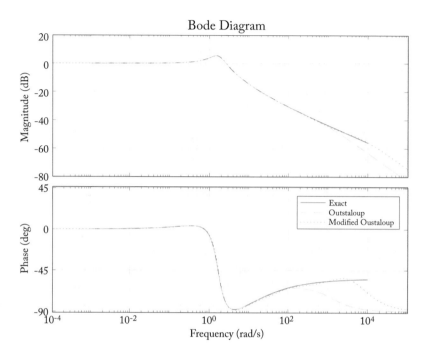

Figure 2.8: Bode plots for Equation (2.5), its Oustaloup, and its modified Oustaloup approximations.

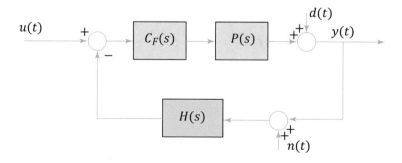

Figure 2.9: Block diagram for a typical fractional PID control system.

deduce the system equation for this closed-loop system, as follows in Equation (2.14), where $Y(s), U(s), N(s)$, and $D(s)$ are Laplace transform of the output, input, noise, and output disturbance signals, respectively:

$$Y(s) = \frac{L(s)}{1+L(s)}U(s) - \frac{L(s)}{1+L(s)}N(s) + \frac{1}{1+L(s)}D(s). \tag{2.14}$$

Correspondingly, the sensitivity transfer function $S(s)$ and the complementary sensitivity transfer function $T(s)$ associated with this system are also defined in Equation (2.15):

$$S(s) = \frac{1}{1+L(s)}, \quad T(s) = \frac{L(s)}{1+L(s)}. \tag{2.15}$$

To design a qualified fractional-order PID controller $C_F(s)$, the following requirements always need to be considered: (1) the system stability can be secured with a phase margin given at the gain crossover frequency; (2) the system is considered robust with the slope of the phase curve approximated to zero in the vicinity of the gain crossover frequency; (3) sensor noise can be rejected by making the gain of $T(s)$ very small at the high frequency; (4) output disturbance can be filtered out by making the gain of $S(s)$ very small at the low frequency; and (5) steady state error can be eliminated by constraining $\lambda > 0$. The specifications for frequency domain design are stated as follows [Kwakernaak and Sivan, 1972, Wang et al., 1995].

(1) Stability requirement:

$$\left|L\left(j\omega_{gc}\right)\right| = 1, \quad \arg\left(L\left(j\omega_{gc}\right)\right) = -\pi + \phi_m. \tag{2.16}$$

(2) Robustness needs:

$$\left|\frac{d}{d\omega}\arg\left(L\left(j\omega\right)\right)\right|_{\omega=\omega_{gc}} \to 0. \tag{2.17}$$

(3) Sensor noise rejection:

$$|T(j\omega)| = \left|\frac{L(j\omega)}{1+L(j\omega)}\right| \le A\,dB = |T\left(j\omega_n\right)|, \quad \forall\,\omega > \omega_n. \tag{2.18}$$

(4) Output disturbance suppression:

$$|S(j\omega)| = \left|\frac{1}{1+L(j\omega)}\right| \le B\,dB = |S\left(j\omega_d\right)|, \quad \forall\,\omega < \omega_d. \tag{2.19}$$

(5) Steady state error elimination:

$$\lambda > 0. \tag{2.20}$$

Parameters K_p, K_i, K_d, λ, and μ in the loop transfer function $L(s)$ need to be solved, with pre-specified system parameters: ω_{gc} is the gain crossover frequency at which the loop gain is unitary; ϕ_m is the phase margin, should always be positive to make the closed loop system stable; ω_n is the lower limit frequency for noise signals need to be rejected; A is a negative decibel value; ω_d is the upper limit frequency for output disturbance signals to be suppressed; and B is a negative decibel value. Based on the frequency domain specification listed above, an optimization can be formulated in Equation (2.21) [Xue, 2017]. Let us denote the to-be-solved parameters are $x = [K_p, K_i, K_d, \lambda, \mu] \in \mathbb{R}^5$:

$$\hat{x} = \arg\min_x \left| \frac{d}{d\omega} \arg(L(j\omega)) \right|_{\omega=\omega_{gc}} \tag{2.21}$$

subject to

$$\left| L(j\omega_{gc}) \right| - 1 = 0$$
$$\arg(L(j\omega_{gc})) + \pi - \phi_m = 0$$
$$20lg \left| \frac{L(j\omega_n)}{1 + L(j\omega_n)} \right| - A \le 0$$
$$20lg \left| \frac{1}{1 + L(j\omega_d)} \right| - B \le 0.$$

Let us use an example to see the performance of a fractional-order PID controller designed through solving the optimization problem in Equation (2.21). At the last part of this section, optimization algorithms will be discussed in more details.

An example is given in Equation (2.22), which means $K = T = 1$. The closed-loop system parameters are also specified: $\omega_{gc} = 0.01$ rad/s, $\phi_m = 60°$, $\omega_n = 10$ rad/s, $\omega_d = 0.001$ rad/s, $A = -20$ dB, and $B = -20$ dB:

$$P(s) = \frac{K}{Ts + 1} = \frac{1}{s + 1}. \tag{2.22}$$

Solving Equation (2.21) leads to the optimal fractional-order PID controller as follows:

$$\widehat{C_F}(s) = \hat{K}_p + \frac{\hat{K}_i}{s^{\hat{\lambda}}} + \hat{K}_d s^{\hat{\mu}} = -13.1177 + \frac{0.0171}{s^{0.9269}} + 13.5452 s^{0.0176}. \tag{2.23}$$

The robustness of the closed-loop system with the optimal fractional-order PID controller can be guaranteed, even when the system structured parameters are identified with uncertainties, for example, parameter K varies in vicinity of $K = 1$. The simulated unit step responses for three sample cases of $K = 0.9$, $K = 1$, and $K = 1.1$ are plotted in Figure 2.10. It shows that this optimal fractional PID controller exhibits good robustness to the unclear modeling parameters of the system dynamics.

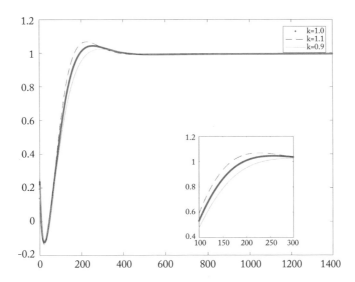

Figure 2.10: Unit step responses with an optimal fractional PID controller with changing parameter K.

2.2 STATE SPACE DESIGN TECHNIQUES

State space representation for a control system plant can simplify the procedures for controller design. This is obvious to be seen when the system dynamics is nonlinear, since its transfer function representation is extremely complex even impossible. In the beginning of this section, we will talk about the state space representations for several typical fractional-order system plants, including the standard-form state space representation, and the general form of the nonlinear state space models. Subsequently, the properties of systems in standard-form state space representations are discussed. The stability criterions presented in the last subsection for the linear fractional-order systems can be reformulated and adapted to the state space representation. Besides, the definitions regarding to the controllability and observability for a given state-space system are also presented. In the last part of this section, the classical pole placement controller is generalized for the cases with fractional-order systems.

2.2.1   STATE SPACE REPRESENTATION

Linear fractional commensurate system with transfer function in Equation (2.2) can also be represented in a state space form, which is given in Equation (2.24). The state space representation in Equation (2.24) is called the standard-form state space representation for a linear fractional system. Note that the fractional operator D^α is not limited to any specific type if the zero initial conditions are assumed. For nonzero initial conditions, we restrict the fractional operator to Caputo's definition, that is, $D^\alpha = {}^C_0 D^\alpha_t$ [Raynaud and Zergainoh, 2000]:

$$D^\alpha x(t) = Ax(t) + Bu(t)$$
$$y(t) = Cx(t) + Du(t). \tag{2.24}$$

As an example, let us convert the transfer function in Equation (2.5) to its state-space form. If we define an intermediate variable $v(t)$ with its Laplace transform $V(s)$ in the transfer function, we will have the following two sub-transfer functions:

$$G(s) = \frac{Y(s)}{V(s)}\frac{V(s)}{U(s)} = \frac{s^{1.8} + 6s^{1.2} + 10s^{0.6} + 12}{2.5s^{2.4} + 5s^{1.8} + 4s^{1.2} + 7.5s^{0.6} + 12} = \frac{Num(s)}{Den(s)}. \tag{2.25}$$

This gives two intermediate transformed relations:

$$\frac{V(s)}{U(s)} = \frac{1}{Den(s)} \qquad \frac{Y(s)}{V(s)} = Num(s). \tag{2.26}$$

By taking the state vector to be $x = \left[D^{1.8}v, D^{1.2}v, D^{0.6}v, v\right]^T$, we get the following results:

$$D^{0.6}x = \begin{bmatrix} D^{2.4}v \\ D^{1.8}v \\ D^{1.2}v \\ D^{0.6}v \end{bmatrix} = \begin{bmatrix} -2 & -1.6 & -3 & -4.8 \\ 1 & 0 & 0 & 0 \\ 0 & 1 & 0 & 0 \\ 0 & 0 & 1 & 0 \end{bmatrix} x + \begin{bmatrix} 1 \\ 0 \\ 0 \\ 0 \end{bmatrix} u$$
$$y = \begin{bmatrix} 1 & 6 & 10 & 12 \end{bmatrix} x. \tag{2.27}$$

Similarly, if we take the state vector to be $x = \left[5D^{1.8}v, 10D^{1.2}v, 20D^{0.6}v, 20v\right]^T$, we will have a new state space representation as follows:

$$D^{0.6}x = \begin{bmatrix} 5D^{2.4}v \\ 10D^{1.8}v \\ 20D^{1.2}v \\ 20D^{0.6}v \end{bmatrix} = \begin{bmatrix} -2 & -0.8 & -0.75 & -1.2 \\ 2 & 0 & 0 & 0 \\ 0 & 2 & 0 & 0 \\ 0 & 0 & 1 & 0 \end{bmatrix} x + \begin{bmatrix} 2 \\ 0 \\ 0 \\ 0 \end{bmatrix} u$$
$$y = \begin{bmatrix} 0.2 & 0.6 & 0.5 & 0.6 \end{bmatrix} x. \tag{2.28}$$

It can be implied from two equivalent state space representations in Equations (2.27) and (2.28) that the state space representation for the given system transfer function is not unique.

Particularly, Equation (2.27) is called the controller canonical form for the state space representation of Equation (2.5).

The solution to fractional LTI system and its proof are stated as follows.

Given a homogeneous nonzero initial value problem defined in Equation (2.29) with $0 < \alpha \leq 1$, the initial state is given as $\boldsymbol{x}(0) = \boldsymbol{x}_0$. The solution to this equation is $\boldsymbol{x}(t) = \Phi(t, 0) \boldsymbol{x}_0$, where $\Phi(t, 0) = E_{\alpha,1}(At^\alpha)$ is called the state transition matrix to the fractional LTI system [Monje et al., 2010]:

$$ {}^C_0 D^\alpha_t \boldsymbol{x}(t) = A\boldsymbol{x}(t). \tag{2.29} $$

Proof. Consider a set of orthogonal bases $\{1, t^\alpha, t^{2\alpha}, \dots\}$, suppose we can express the solution to Equation (2.29) in form of the linear combination of the bases, that is,

$$ \boldsymbol{x}(t) = \boldsymbol{a}_0 + \boldsymbol{a}_1 t^\alpha + \boldsymbol{a}_2 t^{2\alpha} + \cdots \tag{2.30} $$

Taking ${}^C_{t_0} D^\alpha_t$ on both sides of Equation (2.30), we have:

$$ \begin{aligned} {}^C_{t_0} D^\alpha_t \boldsymbol{x}(t) &= \boldsymbol{0} + \boldsymbol{a}_1 \Gamma(\alpha + 1) + \boldsymbol{a}_2 t^\alpha \frac{\Gamma(2\alpha + 1)}{\Gamma(\alpha + 1)} + \boldsymbol{a}_3 t^{2\alpha} \frac{\Gamma(3\alpha + 1)}{\Gamma(2\alpha + 1)} + \cdots \\ &= A\boldsymbol{x}(t). \end{aligned} \tag{2.31} $$

Comparing the coefficients on both sides of Equation (2.31), we have

$$ \boldsymbol{a}_1 \Gamma(\alpha + 1) = A\boldsymbol{a}_0, $$

$$ \vdots $$

$$ \boldsymbol{a}_k \frac{\Gamma(k\alpha + 1)}{\Gamma((k-1)\alpha + 1)} = A\boldsymbol{a}_{k-1}, \tag{2.32} $$

$$ \vdots $$

Since $\boldsymbol{a}_0 = \boldsymbol{x}_0$, we are able to get the formula for all the vector coefficients in Equation (2.30), that is,

$$ \begin{aligned} \boldsymbol{a}_1 &= \frac{A}{\Gamma(\alpha + 1)} \boldsymbol{a}_0 \\ \boldsymbol{a}_2 &= \frac{\Gamma(\alpha + 1)}{\Gamma(2\alpha + 1)} A\boldsymbol{a}_1 = \frac{A^2}{\Gamma(2\alpha + 1)} \boldsymbol{a}_0 \end{aligned} \tag{2.33} $$

$$ \vdots $$

Therefore, Equation (2.30) gives the final solution form as

$$ \boldsymbol{x}(t) = \boldsymbol{x}_0 + \boldsymbol{x}_0 \frac{At^\alpha}{\Gamma(\alpha + 1)} + \boldsymbol{x}_0 \frac{A^2 t^{2\alpha}}{\Gamma(2\alpha + 1)} + \cdots = \boldsymbol{x}_0 E_{\alpha,1}(At^\alpha). \tag{2.34} $$

$\square$

Subsequently, we can move forward to solve the LTI system with control input $u(t)$. Let us consider the nonhomogeneous system in Equation (2.35):

$$_0^C D_t^\alpha x(t) = Ax(t) + Bu(t). \tag{2.35}$$

The solution to Equation (2.35) has a form in Equation (2.36),

$$x(t) = \Phi(t,0)x_0 + \int_0^t \widetilde{\Phi}(t-\tau)Bu(\tau)d\tau, \tag{2.36}$$

where $\widetilde{\Phi}(t)$ is the generalized state transition matrix,

$$\Phi(t,0) = \Phi(t) = \widetilde{\Phi}(t) * \Lambda_{\alpha-1}(t), \tag{2.37}$$

where

$$\Lambda_{\alpha-1}(t) = \begin{cases} \frac{t^{-\alpha}}{\Gamma(1-\alpha)}, & \alpha < 1 \\ \delta(t), & \alpha = 1. \end{cases}$$

Proof. Laplace transforming Equation (2.35) yields

$$s^\alpha X(s) - s^{\alpha-1}x_0 = AX(s) + BU(s). \tag{2.38}$$

By taking the inverse Laplace on Equation (2.38), we get

$$x(t) = \mathcal{L}^{-1}\{X(s)\} = \mathcal{L}^{-1}\left\{(s^\alpha I - A)^{-1}BU(s) + (s^\alpha I - A)^{-1}s^{\alpha-1}x_0\right\}. \tag{2.39}$$

Let us define a help function $\Lambda_{\alpha-1}(t)$,

$$\Lambda_{\alpha-1}(t) = \mathcal{L}^{-1}\{s^{\alpha-1}\} = \begin{cases} \frac{t^{-\alpha}}{\Gamma(1-\alpha)}, & \alpha < 1 \\ \delta(t), & \alpha = 1 \end{cases}. \tag{2.40}$$

By observing Equation (2.39) and the homogeneous solution in Equation (2.34), we have

$$\mathcal{L}^{-1}\left\{(s^\alpha I - A)^{-1}s^{\alpha-1}\right\} = \Phi(t). \tag{2.41}$$

By convolution theorem, it implies

$$\Phi(t) = \mathcal{L}^{-1}\left\{(s^\alpha I - A)^{-1}\right\} * \Lambda_{\alpha-1}(t). \tag{2.42}$$

Denoting $\widetilde{\Phi}(t) = \mathcal{L}^{-1}\left\{(s^\alpha I - A)^{-1}\right\}$, we conclude that

$$x(t) = \Phi(t)x_0 + \int_0^t \widetilde{\Phi}(t-\tau)Bu(\tau)d\tau. \tag{2.43}$$

$\square$

For nonlinear fractional systems, a general form state space representation can also be proposed. Consider a nonlinear commensurate Caputo's fractional differential equation defined in Equation (2.44):

$$\,_{0}^{C}D_t^{n\alpha} y(t) = f\left(t, y(t), \,_{0}^{C}D_t^{\alpha} y(t), \,_{0}^{C}D_t^{2\alpha} y(t), \ldots, \,_{0}^{C}D_t^{(n-1)\alpha} y(t)\right). \tag{2.44}$$

The state variables are introduced to simplify the system dynamics, $x = [x_1, x_2, \ldots, x_n]^T$, such that $x_1 = y(t), x_2 = \,_{0}^{C}D_t^{\alpha} y(t), \ldots, x_n = \,_{0}^{C}D_t^{(n-1)\alpha} y(t)$. Then, we are able to transform Equation (2.44) to a system of equations as follows:

$$\begin{cases} \,_{0}^{C}D_t^{\alpha} x_1(t) = x_2(t) \\ \,_{0}^{C}D_t^{\alpha} x_2(t) = x_3(t) \\ \vdots \\ \,_{0}^{C}D_t^{\alpha} x_{n-1}(t) = x_n(t) \\ \,_{0}^{C}D_t^{\alpha} x_n(t) = f(t, x_1(t), \ldots, x_n(t)). \end{cases} \tag{2.45}$$

Therefore, solving nonlinear fractional equations with form in Equation (2.44) is equivalent to solving a system of equation in Equation (2.46). And this type of equation can be solved using Algorithm 4 discussed in the previous chapter:

$$\,_{0}^{C}D_t^{\alpha} x(t) = \mathcal{F}(t, x(t)). \tag{2.46}$$

2.2.2 STABILITY, CONTROLLABILITY, AND OBSERVABILITY

In the former section, we have given the stability criterion to test the stability of a linear commensurate fractional system given its transfer function. If the system is represented in state-space form, as shown in Equation (2.24), the characteristic polynomial for the fractional system is given will be given in Equation (2.47):

$$\phi(s) = \det\left(s^{\alpha} I - A\right) = 0. \tag{2.47}$$

If we substitute $\zeta = s^{\alpha}$ into Equation (2.47), the transformed characteristic polynomial is given by

$$p(\zeta) = \det\left(\zeta I - A\right) = 0. \tag{2.48}$$

It is straightforward to get the solutions to Equation (2.48) since the solutions are eigenvalues of matrix A. Suppose the eigenvalues of matrix A are denoted by $\zeta_1, \zeta_2, \ldots, \zeta_n$. To test the stability of the system, we only need to evaluate the following inequality:

$$\alpha \frac{\pi}{2} < |\arg(\zeta_i)| < \alpha\pi, \quad \text{for } i = 1, \ldots, n \text{ and } \zeta_i \text{ is genuine poles.} \tag{2.49}$$

For example, we can test the stability of the system described by Equation (2.25), the eigenvalues of its coefficient matrix is given by

$$p(\zeta) = \det(\zeta I - A) = \begin{vmatrix} \zeta + 2 & 1.6 & 3 & 4.8 \\ -1 & \zeta & 0 & 0 \\ 0 & -1 & \zeta & 0 \\ 0 & 0 & -1 & \zeta \end{vmatrix} = 0. \tag{2.50}$$

It is revealed that the poles $\zeta_{1,2} = 0.4802 \pm j1.2739$ and $\zeta_{3,4} = -1.4802 \pm j0.6314$ satisfy the inequality $|\arg(\zeta_{1,2,3,4})| > \frac{\alpha\pi}{2}$ and the genuine poles also satisfy $|\arg(\zeta_{1,2})| < \alpha\pi$. Therefore, we can conclude that this system is stable.

The controllability and observability are two important properties to a given control system. A system is controllable if given an nonzero initial state x_0, there exists a proper control input $u(t)$ such that it can drives the system states back to origin in a finite time interval. That is, $\forall x_0 \neq 0, \exists u(t) \in \mathbb{R}^m (t_0, t_f)$, such that $x(t_f) = 0$ [Balachandran et al., 2013]. For a linear fractional commensurate system as form in Equation (2.24), the system is controllable if and only if the controllability matrix $C = [B, AB, A^2 B, \dots, A^{n-1} B]$ is full rank, that is, rank $(C) = n$. The proof is given as follows.

Proof. Refer to Equation (2.39), we have

$$\begin{aligned} x(t) &= \Phi(t, 0)x_0 + \mathcal{L}^{-1}\left\{(s^\alpha I - A)^{-1} s^{\alpha-1} B s^{1-\alpha} U(s)\right\} \\ &= \Phi(t)x_0 + \Phi(t) * \mathcal{L}^{-1}\left\{B s^{1-\alpha} U(s)\right\} \\ &= \Phi(t)x_0 + \int_0^t \Phi(t - \tau) B\widehat{u}(\tau)d\tau. \end{aligned} \tag{2.51}$$

Note that in Equation (2.51), we denote $\widehat{u}(t) = \mathcal{L}^{-1}\left\{s^{1-\alpha} U(s)\right\}$. If the system is controllable, it means $x(t_f) = 0$:

$$\Phi(t_f) x_0 = -\int_0^{t_f} \Phi(t_f - \tau) B\widehat{u}(\tau)d\tau. \tag{2.52}$$

By Cayley–Hamilton theorem, we can always express the nth power of a matrix as a weighted sum of the lower powers 0 through $n - 1$, that is, for any matrix $A \in \mathbb{R}^{n \times n}$, we have $A^n = \rho_0 I + \rho_1 A + \rho_2 A^2 + \dots + \rho_{n-1} A^{n-1}$. Therefore, we can simplify term $\Phi(t_f - \tau)$ on the right side of Equation (2.52) and we get Equation (2.53). Note that $c_i(\cdot)$ is the coefficient functions obtained by inserting the Cayley–Hamilton A^j, $(j \geq n)$ expansions into Equation (2.53):

$$\Phi\left(t_f - \tau\right) = E_{\alpha,1}\left(A\left(t_f - \tau\right)^{\alpha}\right)$$

$$= \sum_{k=0}^{\infty} \frac{A^k\left(t_f - \tau\right)^{k\alpha}}{\Gamma(k\alpha + 1)} \qquad (2.53)$$

$$= \sum_{i=0}^{n-1} c_i\left(t_f - \tau\right)A^i.$$

Substituting Equation (2.53) into Equation (2.52), we get

$$\Phi\left(t_f\right)x_0 = -\sum_{i=0}^{n-1} A^i B \int_0^{t_f} c_i\left(t_f - \tau\right)\widehat{u}(\tau)d\tau$$

$$= -\left[B, AB, A^2 B, \ldots, A^{n-1}B\right]\begin{bmatrix} \int_0^{t_f} c_0\left(t_f - \tau\right)\widehat{u}(\tau)d\tau \\ \int_0^{t_f} c_1\left(t_f - \tau\right)\widehat{u}(\tau)d\tau \\ \vdots \\ \int_0^{t_f} c_{n-1}\left(t_f - \tau\right)\widehat{u}(\tau)d\tau \end{bmatrix} \qquad (2.54)$$

$$\triangleq -\mathcal{C}\,\widehat{\mathcal{U}}.$$

By observing Equation (2.54), it can be concluded that a proper driving control signal exists if and only if the controllability matrix $\mathcal{C} = \left[B, AB, A^2 B, \ldots, A^{n-1}B\right]$ is full rank. $\square$

Observability is an index for the system transparency. We often refer to a system as observable if in given any initial state x_0, there exists a finite time interval $[t_0, t_f]$, such that we can retrieve the initial state value based on the information of the input $u(t)$ and the output $y(t)$ on the time interval. For a linear fractional commensurate system as in Equation (2.24), we say it is observable if and only if the observability matrix $\mathcal{O}$ is full rank, where the observability matrix is given in Equation (2.55). That is, if we can give rank($\mathcal{O}$) $= n$, then we can assert that the system is observable:

$$\mathcal{O} = \begin{bmatrix} C \\ CA \\ \vdots \\ CA^{n-1} \end{bmatrix}. \qquad (2.55)$$

The proof follows.

Proof. Inserting Equation (2.36) into Equation (2.24), the output is given by

$$y(t) = C\Phi(t)x_0 + C\int_0^t \widetilde{\Phi}(t - \tau)Bu(\tau)d\tau + Du(t). \qquad (2.56)$$

The information of B, C, D, and $\widetilde{\Phi}(t), \boldsymbol{u}(t)$ is given, so that a refined output $\check{\boldsymbol{y}}(t)$ can be obtained

$$\check{\boldsymbol{y}}(t) = \boldsymbol{y}(t) - C \int_0^t \widetilde{\Phi}(t - \tau) B \boldsymbol{u}(\tau) d\tau - D \boldsymbol{u}(t) = C\Phi(t)\boldsymbol{x}_0. \qquad (2.57)$$

Equation (2.53) can be substituted into Equation (2.57) and we obtain

$$\check{\boldsymbol{y}}(t) = C \sum_{i=0}^{n-1} c_i(t) A^i \boldsymbol{x}_0. \qquad (2.58)$$

If we only consider SISO systems, Equation (2.58) can be expressed in matrix form, that is,

$$\check{\boldsymbol{y}}(t) = [c_0(t), c_1(t), \ldots, c_{n-1}(t)] \mathcal{O}\boldsymbol{x}_0 \overset{def}{=} \mathcal{P}(t)\mathcal{O}\boldsymbol{x}_0. \qquad (2.59)$$

Since $\mathcal{P}(t) = [c_0(t), c_1(t), \ldots, c_{n-1}(t)]$ forms a basis for space $\mathbb{R}^n(t)$, the symmetric matrix $\int_0^{t_f} \mathcal{P}^T(t)\mathcal{P}(t)dt$ will be full rank. Multiplying both sides of Equation (2.59) by $\mathcal{P}^T(t)$ and integrating it over the time interval results in

$$\left(\int_0^{t_f} \mathcal{P}^T(t)\mathcal{P}(t)dt\right)^{-1} \int_0^{t_f} \mathcal{P}^T(t)\check{\boldsymbol{y}}(t)dt = \mathcal{O}\boldsymbol{x}_0. \qquad (2.60)$$

By observing Equation (2.60), if the observability matrix $\mathcal{O} \in \mathbb{R}^{n \times n}$ is full rank, the initial state is observable. However, this proof is incomplete since we only consider the system is SISO. For MIMO systems, let us consider the matrix relations as follows:

$$\begin{bmatrix} \boldsymbol{y}(0) - D\boldsymbol{u}(0) \\ {}_0^C D_t^\alpha \boldsymbol{y}(0) - D {}_0^C D_t^\alpha \boldsymbol{u}(0) - CB\boldsymbol{u}(0) \\ {}_0^C D_t^{2\alpha} \boldsymbol{y}(0) - D {}_0^C D_t^{2\alpha} \boldsymbol{u}(0) - CB {}_0^C D_t^\alpha \boldsymbol{u}(0) - CAB\boldsymbol{u}(0) \\ \vdots \\ {}_0^C D_t^{(n-1)\alpha} \boldsymbol{y}(0) - D {}_0^C D_t^{(n-1)\alpha} \boldsymbol{u}(0) - \ldots - CA^{n-2}B\boldsymbol{u}(0) \end{bmatrix} = \begin{bmatrix} C \\ CA \\ CA^2 \\ \vdots \\ CA^{n-1} \end{bmatrix} \boldsymbol{x}_0 = \mathcal{O}\boldsymbol{x}_0. \qquad (2.61)$$

As Equation (2.61) shows, the system input is $\boldsymbol{u}(t) \in \mathbb{R}^m$ and output is $\boldsymbol{y}(t) \in \mathbb{R}^p$. We see that the observability matrix must be full rank to solve for the initial state $\boldsymbol{x}_0$. This completes the proof. $\qquad \square$

2.2.3 POLE PLACEMENT IN FRACTIONAL-ORDER STATE SPACE CONTROL

Pole placement acts as a "Hello World!" role among state space controller design strategies. The pole placement method in fractional case is similar to that in the integer order case. First of all, we need to extract out the full states and feedback them with associated gains. Second, a new compound coefficient matrix pairs can be formed. By changing the associated gains of the full state feedbacks, the poles of the closed-loop system can be placed to a desired position. The new pole positions will determine the control performance. The control objectives such as stability, robustness, noise rejection, and disturbance elimination can be achieved if appropriate pole positions are selected. The system block diagram for fractional-order full-state feedback control is shown in Figure 2.11.

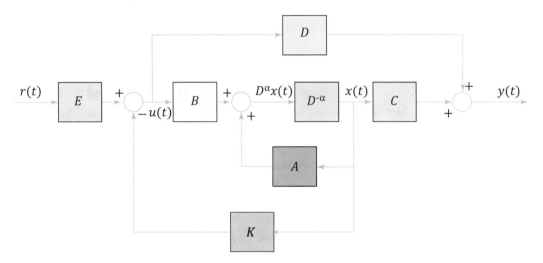

Figure 2.11: System block diagram for a typical fractional-order full-state feedback system.

In this full-state feedback system, we refer to $r(t)$ as an exogenous input. Then state space representation of this closed-loop system is given in (2.62):

$$D^\alpha x(t) = (A - BK)x(t) + BEr(t)$$
$$y(t) = (C - DK)x(t) + DEr(t).$$

$$(2.62)$$

Suppose that desired poles are given at $\zeta_1, \zeta_2, \ldots, \zeta_n$, which should satisfy the stability criterion $\|\arg(\zeta_i)\| > \alpha \frac{\pi}{2}$. Given the desired pole positions, the feedback gain matrix K can be obtained by solving the following equation:

$$\det(\zeta I - (A - BK)) = \prod_{i=1}^{n} (\zeta - \zeta_i).$$

$$(2.63)$$

2.3 OPTIMIZATIONS OF FRACTIONAL-ORDER CONTROLLERS

Control engineering problems are essentially optimization problems. Given constrained conditions and a specified cost function, a group of parameters may be solved to minimize the cost function. We say this group of parameters is optimal in regard to the specific cost function. Since our emphasis is placed on the controller design and the parameters of a controller are usually constrained within a compact real set formed by the constrained conditions, it shows that controller optimization problems are mainly constrained optimization problems. The unconstrained minimization problem has a general form formulated in Equation (2.64), where $G \in \mathbb{R}^n$ is the feasible set. The best solution would be vectors which can minimize the cost function $f(x)$ over the feasible set:

$$\begin{aligned} \text{minimize} \quad & f(x) \\ \text{subject to} \quad & x \in G. \end{aligned} \tag{2.64}$$

To solve this type of problems, many algorithms have been developed by researchers. For example, if the cost function is continuously differentiable and the goal is to find the local minimizers, the algorithms like Gradient methods, Newton–Rahpson's method, Conjugate Direction method, Conjugate Gradient method, and Quasi-Newton methods can be applied. These methods involve steps to find the minimal solution along a one-dimensional search line, and algorithms such as Golden Section, Fibonacci, Bisection, Newton's, or Secant methods can be applied [Chong and Zak, 2013]. In order to solve for global optimizers under unconstrained conditions, global search algorithms like genetic algorithms (GA), and particle swarm optimization (PSO) algorithms can be applied. The constrained optimization problems also hold a general form in Equation (2.65). Algorithms based on first-order sufficiency condition (FOSC) or second-order sufficiency condition (SOSC) was also discussed in many optimization sources:

$$\begin{aligned} \text{minimize} \quad & f(x) \\ \text{subject to} \quad & h(x) = 0 \\ & g(x) \leq 0. \end{aligned} \tag{2.65}$$

Typical as it is for optimal control problems, the cost function $f(x)$ is usually formulated to satisfy the specified control objectives, for example, the integral of squared error J_1, the integral of absolute error J_2, the integral of time weighted absolute error J_3, etc. Similarly, the cost function can be given as a weighted sum of multiple control performance indicator variables. For the specific applications in the next chapter, we will discuss how to design an effective cost function in more details:

$$J_1 = \int_0^\infty e^2(t)dt, \quad J_2 = \int_0^\infty |e(t)|\, dt, \quad J_3 = \int_0^\infty t\, |e(t)|\, dt. \tag{2.66}$$

Let us consider again the system plant in Equation (2.5). We can solve for a fractional-order PID controller by minimizing the cost function over a specified feasible set. The new minimization problem is formulated in Equation (2.67), where G is a set such that $K_p, K_i, K_d \in [0, 2]$, and $\lambda, \mu \in [0, 20]$. Note that the initial range of the feasible set is given empirically:

$$\left[\tilde{K}_p, \tilde{K}_i, \tilde{K}_d, \tilde{\lambda}, \tilde{\mu} \right] = \arg \min_{K \in G} J_3(K_p, K_i, K_d, \lambda, \mu). \tag{2.67}$$

The optimal parameters of the fractional-order PID controller is computed and the final form of the optimal fractional-order PID controller is given in Equation (2.68):

$$C_F(s) = 1.9368 \times 10^{-6} + \frac{19.9999}{s^{1.0228}} + 8.8745s^{0.5427}. \tag{2.68}$$

The open-loop unit step response and the closed-loop unit step response under this controller is shown in Figure 2.12. The closed-loop system response performance is shown much better than the open loop.

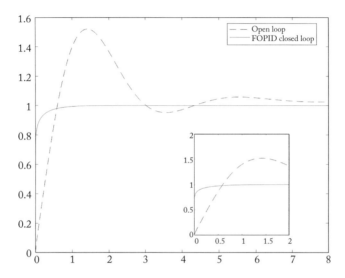

Figure 2.12: Unit step response for open-loop system and closed-loop system.

CHAPTER 3

Control Applications in Engineering and Technology

3.1 FRACTIONAL-ORDER ADAPTIVE CONTROL OF MACHINING CHATTER

When the cutting parameters of machining process fall into the unstable conditions, chatter occurs at any trivial disturbances. Essentially, machining chatter is the self-excited vibration with time delays, and it has harmful effects on cutting tool life and workpiece surface finish. However, significantly changing the cutting parameters is not realistic in the industrial operations due to the requirements of cycle time and production throughputs.

In search of prompt machining chatter suppression, the adaptive control strategy is employed. In this section, a fractional-order adaptive control is derived from Lyapunov function without dramatically changing the cutting conditions. The fractional-order adaptive control is attested effective in numerical simulation.

3.1.1 MODELING OF THE SINGLE-DELAY REGENERATIVE MACHINING CHATTER

For simplicity and without loss of generality, the regenerative machining chatter in single point cutting process is modeled as a nonlinear differential-difference equation with single delay:

$$
\begin{cases}
\frac{d^2 x}{dt^2} + 2\zeta\omega_n \frac{dx(t)}{dt} + \omega_n^2 x(t) = -\frac{\omega_n^2}{K}\triangle F(t)\cos\alpha \\
\triangle F(t) = Pw\left\{s^\rho(t) - s_0^\rho + \frac{\rho s_0^{\rho-1} C}{N}\frac{dx}{dt}\right\} \\
s(t) = \begin{cases} x(t) - x(t-T) + s_0, & \text{if } s(t) > 0 \\ 0, & \text{otherwise,} \end{cases}
\end{cases}
\tag{3.1}
$$

where ω_n is natural frequency, ζ is damping ratio, K is modal stiffness, $x(t)$ is relative displacement between the tool and the workpiece in the direction normal to the cutting surface, $\triangle F(t)$ is dynamic force, P is force coefficient, w is depth of cut, C is penetration coefficient, $s(t)$ is instantaneous chip thickness, s_0 is normal chip thickness, N is spindle speed, ρ is force exponential of value between $(1, 2)$, T is period of workpiece revolution, and α is angle between cutting force $F(t)$ and displacement $x(t)$.

The nonlinear part $s^\rho(t) - s_0^\rho = [x(t) - x(t - T) + s_0]^\rho - s_0^\rho$ could be first-order approximated, and the linear approximation of the above chatter model is as follows:

$$
\begin{cases}
\frac{d^2x}{dt^2} + a_1 \frac{dx(t)}{dt} + \left(\omega_n^2 + a_2\right) x(t) = a_2 x(t - T) \\
a_1 = 2\zeta\omega_n + \frac{\omega_n^2}{K} P w \rho s_0^{\rho-1} C \frac{1}{N} \cos\alpha \\
a_2 = \frac{\omega_n^2}{K} P w \rho s_0^{\rho-1} \cos\alpha.
\end{cases}
\tag{3.2}
$$

3.1.2   ADAPTIVE CONTROL LAW DESIGN

Given the reference model being a stable machining system, which could be described by a regular order differential difference equation as in Equation (3.3):

$$
\dot{Y}(t) = A\,Y(t) + B\,Y(t - T),
\tag{3.3}
$$

where $Y(t) \in \mathbb{R}^{n\times 1}$ ($n = 2$ in this case). And the corresponding unstable machining system is represented by Equation (3.4):

$$
\dot{X}(t) = (A + \delta A)X(t) + (B + \delta B)X(t - T - h),
\tag{3.4}
$$

where $X(t) \in \mathbb{R}^{n\times 1}$. For the case of single point turning process, in Equations (3.3) and (3.4), X and Y are vector whose elements are the chatter displacement and its velocity; the time delay is $T = \frac{60}{N} - h$, where h is the variable time delay, and δA and δB are the parameter deviations from the matrices $A \in \mathbb{R}^{n\times n}$ and $B \in \mathbb{R}^{n\times n}$:

$$
A = \begin{bmatrix} 0 & 1 \\ -\omega_n^2 - a_2 & -a_1 \end{bmatrix}, \quad
B = \begin{bmatrix} 0 & 0 \\ a_2 & 0 \end{bmatrix}
\tag{3.5}
$$

$$
\delta A = -\delta B = \begin{bmatrix} 0 & 0 \\ -\delta a_2 & 0 \end{bmatrix}.
\tag{3.6}
$$

The stability conditions for Equation (3.3) are given as follows [Lee and Dianat, 1981]:

$$
P\left[A + Q(0)\right] + \left[A + Q(0)\right]' P + R = 0
\tag{3.7}
$$

$$
\dot{Q}(t) = [A + Q(0)]Q(t), \quad t \in [0, T].
\tag{3.8}
$$

The error state equation is $\varepsilon(t) = X(t) - Y(t)$. The error state equation can be approximated by

$$
\dot{\varepsilon}(t) \approx A\varepsilon(t) + B\varepsilon(t - T) + \phi Z.
\tag{3.9}
$$

Since the variable h is trivial, and the difference of $X(t - T + h)$ and $X(t - T)$ can be assessed by $h\frac{dX(t-T)}{dt}$.

In Equation (3.9), $\phi \in \mathbb{R}^{n\times 3n}$ is the adjustable parameter matrix,

$$
\phi = \begin{bmatrix} 0 & 0 & 0 & 0 & 0 & 0 \\ -\delta a_2 & 0 & \delta a_2 & 0 & a_2 h & 0 \end{bmatrix}
\tag{3.10}
$$

and the augmented state vector $Z \in \mathbb{R}^{3n \times 1}$

$$Z = \begin{bmatrix} X(t) \\ X(t - T + h) \\ \dot{X}(t - h) \end{bmatrix}. \tag{3.11}$$

The asymptotic stability in the error space is achieved when all the error space elements $\varepsilon(t)$, $\delta A(t)$, $\delta B(t)$, and $h(t)$ become zero in steady states. It is proved that for this single-delay machining system is asymptotically stable as $\lim_{t \to \infty} \varepsilon(t) = 0$ [Zhang, 1996]. Its control law is presented as follows:

$$\delta \dot{f}(t) = -\frac{\lambda_2 s_0 x(t - T + h)}{a_2(\rho - 1) \sin \kappa} \\ \cdot [P_{12}(\varepsilon_1 + Q_{11} \cdot \varepsilon_1 + Q_{12} \cdot \varepsilon_2) + P_{22}(\varepsilon_2 + Q_{21} \cdot \varepsilon_1 + Q_{22} \cdot \varepsilon_2)] \tag{3.12}$$

$$\dot{h} = -\frac{\lambda_3 \dot{x}(t - T)}{a_2} \\ \cdot [P_{12}(\varepsilon_1 + Q_{11} \cdot \varepsilon_1 + Q_{12} \cdot \varepsilon_2) + P_{22}(\varepsilon_2 + Q_{21} \cdot \varepsilon_1 + Q_{22} \cdot \varepsilon_2)], \tag{3.13}$$

where κ is the tool cutting edge angle, $P = [P_{ii}]$ $(i = 1, \ldots, n)$ is the positive definite Hermitian, and $Q(t) \in \mathbb{R}^{n \times n}$ is a characteristic matrix being continuous and differentiable over $[0, T]$ and zero elsewhere.

In order to design more advanced control for the machining chatter, the advantage of fractional calculus is taken. It is desirable that the control law with the chatter states $x(t)$ and $\dot{x}(t)$ can include the information of acceleration $\ddot{x}(t)$. The fractional-order control law can serve this purpose with the fractional order less than n ($n = 2$ in this case).

In the fractional order state space, the stable differential reference model with a single time delay at order β (a real number) is

$$^C_0 D_t^\beta Y(t) = \tilde{A} Y(t) + \tilde{B} Y(t - T), \tag{3.14}$$

and its corresponding unstable fractional-order model is

$$^C_0 D_t^\beta X(t) = (\tilde{A} + \delta \tilde{A}) X(t) + (\tilde{B} + \delta \tilde{B}) X(t - T - h). \tag{3.15}$$

In Equations (3.14) and (3.15), the operator $^C_0 D_t^\beta$ is the fractional derivative of order β with respect to t, starting at point $t = 0$ [Caputo, 1969]:

$$^C_0 D_t^\beta f(t) = \frac{1}{\Gamma(\beta - n)} \int_0^t \frac{f^{(n)}(\tau) d\tau}{(t - \tau)^{\beta + 1 - n}}, \quad (n - 1 < \beta < n). \tag{3.16}$$

The error space of this fractional order system is:

$$\begin{aligned}
{}_0^C D_t^\beta \varepsilon(t) &= {}_0^C D_t^\beta \left[X(t) - Y(t) \right] \\
&= \tilde{A}\varepsilon(t) + \tilde{B}\varepsilon(t-T) + \delta\tilde{A}x(t) \\
&\quad + \tilde{B}\left[x(t-T+h) - x(t-T) \right] + \delta\tilde{B}x(t-T+h) \\
&\approx \tilde{A}\varepsilon(t) + \tilde{B}\varepsilon(t-T) + \tilde{\phi}\tilde{Z},
\end{aligned} \tag{3.17}$$

where $\tilde{A}$, $\tilde{B}$, and $\tilde{\phi}$ are matrices which have same dimensions as that of Equations (3.5), (3.6), and (3.10) and are with scalar elements, and the augmented state vector $\tilde{Z} \in \mathbb{R}^{3n \times 1}$:

$$\tilde{Z} = \begin{bmatrix} X(t) \\ X(t-T+h) \\ {}_0^C D_t^\beta X(t-h) \end{bmatrix}. \tag{3.18}$$

This fractional-order error space contains the information of position, velocity, and acceleration of the self-excited vibration during the machining chatter. For the nonlinear differential-difference model of machining chatter, due to the similarity between these two error spaces, a fractional-order control law can be derived with different scalar parameters and the fractional derivatives. As a matter of fact, the ordinary order control law expressed in Equations (3.12) and (3.13) can be viewed as a special case $\beta = 1$ of the fractional-order control law.

Wan, Zhang and French [2015] also presented another approach, utilizing the bias between the fractional state and the integer state of the machining chatter:

$$\begin{aligned}
Bias &= {}_0^C D_t^\beta X(t) - \dot{X}(t) \\
&= \left(\tilde{A} + \delta\tilde{A} - A - \delta A \right) \cdot X(t) + \left(\tilde{B} + \delta\tilde{B} - B - \delta B \right) X(t-T).
\end{aligned} \tag{3.19}$$

At given disturbances, it converges to zero when the machining process is stable, and it diverges when the machining chatter occurs. Therefore, this bias could be an index for the onset of machining chatter. As defined by the Caputo's fractional derivative, ${}_0^C D_t^\beta f(t)$ is equivalent to the linear combination of $F\left(f^{(n)}(\tau) \right)$, and $F(0) = 0$. As the machining chatter is fully suppressed, the states $x(t) = \dot{x}(t) = \ddot{x}(t) = 0$.

The fractional integrators of the spindle speed and feed rate will bring enhanced controls to the machining process. Suppressing the machining chatter in the onset stage is always easier and better, thus it is desirable to limit the bias to small scales by adjusting the fractional order $\beta(t)$. For the machining chatter control, dimension $n = 2$, given the error space $\lim_{t \to \infty} \varepsilon(t) = 0$, one could tune the fractional order $\beta(t)$ to satisfy the following condition: $\beta(t_0) \leq 2, \beta(t_{stb}) = 1, \beta'(t) < 0, t \in [t_0, t_{stb}]$, where t_0 is the time the chatter onsets and t_{stb} is the time the machining process is stabilized. This condition retains the asymptotically stability of the adaptive control, since it converges to the ordinary order within a finite time window.

The new control laws of the spindle speed and feed rate are expressed as follows:

$$_0^C D_t^\beta \delta f(t) = -\frac{\lambda_2 s_0 x(t - T + h)}{a_2(\rho - 1)\sin\kappa} \cdot G\left(\tilde{P}, \tilde{Q}, \tilde{\varepsilon}, Bias, \beta\right) \tag{3.20}$$

$$_0^C D_t^\beta h(t) = -\dot{x}(t - T)\frac{\lambda_3}{a_2} \cdot H\left(\tilde{P}, \tilde{Q}, \tilde{\varepsilon}, Bias, \beta\right). \tag{3.21}$$

In these two control laws, $G\left(\tilde{P}, \tilde{Q}, \tilde{\varepsilon}, Bias, \beta\right)$ and $H\left(\tilde{P}, \tilde{Q}, \tilde{\varepsilon}, Bias, \beta\right)$ are scalar parameters determined by $\tilde{P}, \tilde{Q}, \tilde{\varepsilon}, Bias$, and β. Although the control is a biased at early stage, the tradeoff is faster suppression and the augmented stability with higher precision.

For simplicity and feasibility for the industrial applications, a stepwise downward tuning with the fractional-order β adjustments is used:

$$\begin{cases} \text{for}: \ t_0 < t_1 < \cdots < t_n < t_{stb} \\ \text{set}: \ 2 \geq (t_0) > (t_1) > \cdots (t_n) > (t_{stb}) = 1. \end{cases} \tag{3.22}$$

The numerical simulation shows the adjustable fractional-order adaptive control is superior, and it achieves a quick suppression of the chatter and avoids over adjustment on the spindle speed.

3.1.3 NUMERICAL SIMULATION

In this section, using the integer order adaptive control (its control laws are given in Equations (3.12) and (3.13)) as a baseline, other fixed fractional-order adaptive controls and adjustable fractional-order adaptive control are numerically simulated and compared to evaluate their control effectiveness.

In Figure 3.1, the fixed fractional control simulations of relative displacement (Figure 3.1a) and time delay (Figure 3.1b) of the machining chatter (with $\beta = 1.3, 1.6$, or 1.9, respectively) are assessed against the integer order adaptive control (with $\beta = 1.0$). The cutting parameters for numerical simulation are: depth of cut (unstable) 5 mm, depth of cut (stable) 2 mm, initial spindle speed of 450 rpm, initial feed rate of 0.1 mm/s, damping ratio of 0.038, and natural frequency of 85 Hz.

It is observed that the three fixed fractional-order adaptive controls suppress the machining chatter quicker than the integer order adaptive control, and the case of $\beta = 1.9$ is the quickest. However, a disadvantage of these fixed fractional order adaptive controls is that during the machining chatter suppression the spindle speed adjustments are much larger than that of the integer order adaptive control. This is reflected in curves of Figure 3.1b.

The machining chatter's first-order approximation of the relative displacement is:

$$x(t) = \text{Real}\left\{A(t)e^{j\varphi(t)}\right\}, \tag{3.23}$$

where $A(t)$ is the chatter amplitude, $\varphi(t) = \omega(t) \cdot t$ is the phase angle, and $\omega(t) = d\varphi(t)/dt$ is the frequency.

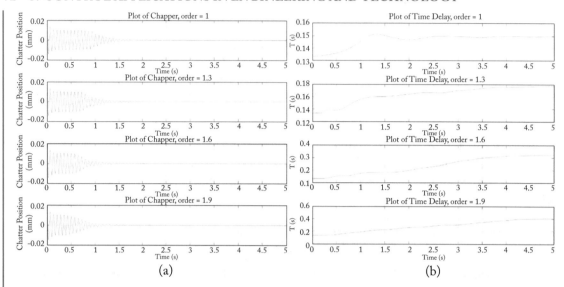

Figure 3.1: Integer-order adaptive control vs. fixed fractional-order adaptive control: (a) chatter and (b) time-delay.

The net work done by the dynamic cutting force in a vibration cycle during the machining process is derived as:

$$
\begin{aligned}
E = PwA^2\rho\cos\alpha \Bigg\{ &-s_0^{\rho-1}\pi C\frac{\omega}{N} - s_0^{\rho-1}\pi\sin\varphi + \frac{3.77}{2}A(\rho-1)s_0^{\rho-2}(1-\cos\varphi) \\
&+ \frac{1}{6}A^2(\rho-1)(\rho-2)s_0^{\rho-3}\pi \\
&\left[-\frac{3}{4}\sin\varphi + \frac{6}{4}\sin\varphi\cos\varphi - \frac{3}{4}\sin\varphi\cos^2\varphi - \frac{3}{4}\sin^3\varphi \right] \Bigg\}.
\end{aligned}
\tag{3.24}
$$

For the effective chatter suppression, the net work of a chatter cycle done by the dynamic cutting force decreases, so that the self-excited vibration dies out due to less and less energy supplied into it. This is verified by the numerical simulations of the integer order adaptive control and fixed fractional order adaptive controls, which are illustrated in Figure 3.2.

Therefore, the adjustable fractional-order adaptive control is employed to achieve enhanced suppression effect. The bias issue of the machining chatter control is directly related to the net work of self-excited vibration cycle. The adjustable fractional order adaptive control optimized via selecting the fractional order β stepwise is simulated and illustrated in Figure 3.3.

It is demonstrated that the adjustable fractional order adaptive control is effective for machining chatter suppression without dramatically changing the cutting conditions. This fractional-order approach is based on the control laws derived from Lyapunov function for the

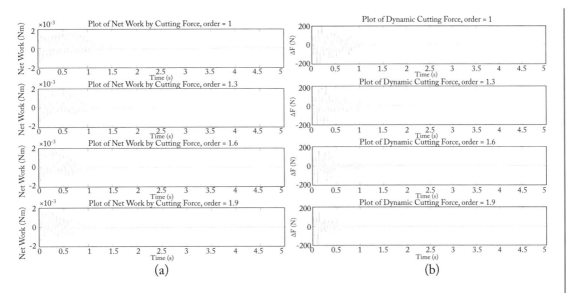

Figure 3.2: Net work in self-excited vibration cycle vs. dynamic cutting force: (a) chatter and (b) time delay.

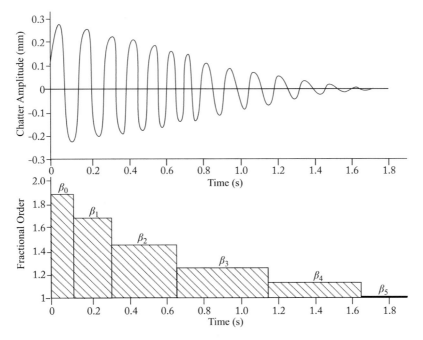

Figure 3.3: The stepwise adjustment of fractional order β and its effectiveness (Wan, Zhang and French [2015]).

linearized single delay differential-difference equation, and tuned with the internal energy analysis of the machining process that the single delay differential-difference equation describes. Therefore, the effectiveness of the adjustable fractional-order adaptive control is guaranteed. It is a promising control for the machining chatter control and its prediction at the stage of chatter onset.

3.2   FRACTIONAL-ORDER PID DESIGN FOR NONLINEAR MOTION CONTROL OF A MULTILINK ROBOT

Multilink robots, such as Adept 550, widely used in manufacturing industry, have always demanded better motion control since their advent. Mathematically, they are typically high-dimensional systems with nonlinearities, uncertain disturbances, and non-unique characteristic functions. Multilink robots often employ PID controller, due to its simplicity and feasibility, for their motion controls. However, for precision controls of high-dimension systems, PIDs performance is outshone in comparison with that of other advanced controllers, which normally are designed through complicated control theories. Utilizing the advantages of this fractional calculus, this section focuses on developing the fractional-order PID controllers that retain the unique simplicity, at the same time serve the purpose of precision control for higher dimension systems.

The dynamic model of the multilink robot is first established. It provides the foundation for the control law design, and the objective function of the optimization design of the subjected fractional order control system. The closed-loop transaction function in frequency domain of the studied fractional system has been developed and the paper proceeds by the study of controllability, observability, and robustly satiability. It is demonstrated that the fractional order PID controller could significantly increase the stability of trajectory tracking of the multilink robot system, and therefore brings superior control performance in terms of accuracy.

3.2.1   SYSTEM DYNAMICS MODELING OF A MULTILINK ROBOT

A multilink robot is illustrated in Figure 3.4. This multilink robot is composed of two rigid bodies: inner ($i = 1$) and outer ($i = 2$) links, which are linked by revolving joints, and have the central mass $m_i (i = 1, 2)$. The gripper and load are symbolized as a mass m_3. Without loss of generality, the gripper angle adjustment is simplified by assuming the wrist rotation angle is zero, so that the dynamics modeling for the trajectory tracking can be developed [Zhang, 2010].

Applying Denavit–Hartenberg (D-H) coordinates with the parameters summarized in Table 3.1, the gripper's horizontal position (P_x, P_y) is expressed as:

$$\begin{aligned} P_x &= L_1 \cos \beta_1 + L_2 \cos \beta_2 \\ P_y &= L_1 \sin \beta_1 + L_2 \sin \beta_2, \end{aligned} \tag{3.25}$$

where $\beta_i (i = 1, 2)$ are the angular positions of the actuating motors, and $\theta_i (i = 1, 2)$ are the angles about the two parallel Z axes. Their relationships are as follows: $\beta_1 = \theta_1, \beta_2 = \theta_1 + \theta_2$.

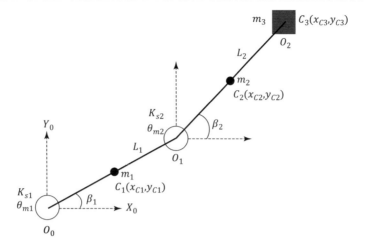

Figure 3.4: A multilink robot.

Table 3.1: D-H parameters

Link	Li	i	di	θ_i
Inner	$L1$	0	0	θ_1
Outer	$L2$	0	0	θ_2

The motor angular positions (β_1, β_2) are expressed as

$$\beta_1 = 2\tan^{-1}\left(\frac{P_y \pm \sqrt{P_x^2 + P_y^2 - R_1}}{P_x + R_1}\right)$$

$$\beta_2 = 2\tan^{-1}\left(\frac{P_y \pm \sqrt{P_x^2 + P_y^2 - R_2}}{P_x + R_2}\right),$$

(3.26)

where

$$R_1 = \frac{P_x^2 + P_y^2 + L_1^2 - L_2^2}{2L_1}$$

$$R_2 = \frac{P_x^2 + P_y^2 + L_2^2 - L_1^2}{2L_2}.$$

And the forward velocity and acceleration of the gripper are as follows:

$$\begin{pmatrix} \dot{P}_x \\ \dot{P}_y \end{pmatrix} = \begin{bmatrix} -L_1 \sin \beta_1 & -L_2 \sin \beta_2 \\ L_1 \cos \beta_1 & L_2 \cos \beta_2 \end{bmatrix} \begin{pmatrix} \dot{\beta}_1 \\ \dot{\beta}_2 \end{pmatrix}. \tag{3.27}$$

$$\begin{pmatrix} \ddot{P}_x \\ \ddot{P}_y \end{pmatrix} = \begin{bmatrix} -L_1 \sin \beta_1 & -L_2 \sin \beta_2 \\ L_1 \cos \beta_1 & L_2 \cos \beta_2 \end{bmatrix} \begin{pmatrix} \ddot{\beta}_1 \\ \ddot{\beta}_2 \end{pmatrix} + \begin{bmatrix} -L_1 \cos \beta_1 & -L_2 \cos \beta_2 \\ -L_1 \sin \beta_1 & -L_2 \sin \beta_2 \end{bmatrix} \begin{pmatrix} \dot{\beta}_1 \\ \dot{\beta}_2 \end{pmatrix}. \tag{3.28}$$

For modeling the robot dynamics, Lagrange method is employed. The Lagrange function L is the difference of the total kinematic energy T and the total potential energy V, i.e.,

$$L = T - V, \tag{3.29}$$

where

$$\begin{aligned} T = {} & \frac{1}{2}m_1 \left(\frac{1}{4}L_1^2 \dot{\beta}_1^2 \right) + \frac{1}{2}m_2 \left(L_1^2 \dot{\beta}_1^2 + L_1 L_2 \dot{\beta}_1 \dot{\beta}_2 \cos(\beta_2 - \beta_1) + \frac{1}{4}L_2^2 \dot{\beta}_2^2 \right) \\ & + \frac{1}{2}m_3 \left(L_1^2 \dot{\beta}_1^2 + 2L_1 L_2 \dot{\beta}_1 \dot{\beta}_2 \cos(\beta_2 - \beta_1) + L_2^2 \dot{\beta}_2^2 \right) + \frac{1}{2} \left(\frac{1}{3}m_1 L_1^2 \right) \dot{\beta}_1^2 \\ & + \frac{1}{2} \left(\frac{1}{3}m_2 L_2^2 \right) \dot{\beta}_1^2 \end{aligned}$$

$$\begin{aligned} V = {} & \left(\frac{1}{2}m_1 g L_1 + m_2 g L_1 + m_3 g L_1 \right) \sin \beta_1 + \left(\frac{1}{2}m_2 g L_2 + m_3 g L_2 \right) \sin \beta_2 \\ & + \frac{1}{2}k_{s1} \left(r\theta_{m1} - \beta_1 \right)^2 + \frac{1}{2}k_{s2} \left(r\theta_{m2} - \beta_2 \right)^2. \end{aligned}$$

Using

$$\frac{d}{dt} \left(\frac{\partial T}{\partial \dot{\beta}_1} \right) - \frac{\partial T}{\partial \beta_1} + \frac{\partial V}{\partial \beta_1} = \tau_1$$

$$\frac{d}{dt} \left(\frac{\partial T}{\partial \dot{\beta}_2} \right) - \frac{\partial T}{\partial \beta_2} + \frac{\partial V}{\partial \beta_2} = \tau_2.$$

The system dynamics of the multilink robot is derived as follows:

$$D(\beta)\ddot{\beta} + H\left(\beta, \dot{\beta}\right)\ddot{\beta} + G(\beta) = \tau + \tau_{damping}, \tag{3.30}$$

where

$$D(\beta) = \begin{bmatrix} \left(\frac{7}{12}m_1 + m_2 + m_3 \right) L_1^2 & \left(\frac{1}{2}m_2 + m_3 \right) L_1 L_2 \cos (\beta_2 - \beta_1) \\ \left(\frac{1}{2}m_2 + m_3 \right) L_1 L_2 \cos (\beta_2 - \beta_1) & \left(\frac{7}{12}m_2 + m_3 \right) L_2^2 \end{bmatrix}$$

$$H\left(\beta, \dot{\beta}\right) = \begin{bmatrix} 0 & -\left(\frac{1}{2}m_2 + m_3\right) L_1 L_2 \sin\left(\beta_2 - \beta_1\right) \dot{\beta}_2 \\ \left(\frac{1}{2}m_2 + m_3\right) L_1 L_2 \sin\left(\beta_2 - \beta_1\right) \dot{\beta}_1 & 0 \end{bmatrix}$$

$$G(\beta) = \begin{pmatrix} \left(\frac{1}{2}m_1 + m_2 + m_3\right) g L_1 \cos\beta_1 - k_{s1}\left(r\theta_{m1} - \beta_1\right) \\ \left(\frac{1}{2}m_2 + m_3\right) g L_2 \cos\beta_2 - k_{s2}\left(r\theta_{m2} - \beta_2\right) \end{pmatrix}$$

$$\tau = \begin{pmatrix} \tau_1 \\ \tau_2 \end{pmatrix}$$

$$\tau_{damping} = \begin{pmatrix} \tau_{damping1} \\ \tau_{damping2} \end{pmatrix} = \begin{bmatrix} -C_1 & 0 \\ 0 & -C_2 \end{bmatrix} \begin{pmatrix} \dot{\beta}_1 \\ \dot{\beta}_2 \end{pmatrix},$$

where τ is torque vector composed of the torque elements, and $C_i (i = 1, 2)$ are the damping coefficients contributing to the viscous torques at the joints.

Assuming that the identical motors are used to drive the inner and outer links, including the electric motor dynamics into the dynamics of the multilink robot yields the system dynamic model:

$$\sum_{j=1}^{n} d_{jk}(\beta)\ddot{\beta}_j + \sum_{i,j=1}^{n} h_{ijk}(\beta)\dot{\beta}_i\dot{\beta}_j + g_k(\beta) = \tau_k - C_k\dot{\beta}_k \tag{3.31}$$

$$J_{m,k}\ddot{\theta}_{mk} + \left(B_{m,k} + K_{b,k}\frac{K_{m,k}}{R_k}\right)\dot{\theta}_{mk} = \frac{K_{m,k}}{R_k}V_k - \tau_{m,k}, \quad k = 1, 2, \tag{3.32}$$

where $J_{m,k}$ is the rotor inertia, $B_{m,k}$ is the viscosity, $K_{b,k}$ and $K_{m,k}$ are motor cosntants, and R_k is the armature resistance.

Utilizing the relationship between the link angle β_i and motor rotating angle $\theta_{m,i}$: $\beta_i = r\theta_{m,i}$, and the relationship between the link torque and the motor torque: $\tau_{m,i} = r\tau_i (i = 1, 2)$, where r is the gear ratio of the electric motor, Equations (3.31) and (3.32) can be combined into a single one:

$$J_{eff,k}\ddot{\theta}_{mk} + B_{eff,k}\dot{\theta}_{mk} + C_k\theta_{mk} = KV_k - rd_k, \quad k = 1, 2, \tag{3.33}$$

where $J_{eff,k}$ is the effective inertia, $B_{eff,k}$ is the effective viscosity, and $d_k (k = 1, 2)$ are consisted of nonlinear terms, treated as the disturbances to the system, expressed as:

$$d_1 = \left(\frac{1}{2}m_2 + m_3\right) L_1 L_2 \cos\left(\beta_2 - \beta_1\right) \ddot{\beta}_2 - \left(\frac{1}{2}m_2 + m_3\right) L_1 L_2 \sin\left(\beta_2 - \beta_1\right) \dot{\beta}_2^2$$
$$+ \left(\frac{1}{2}m_1 + m_2 + m_3\right) g L_1 \cos\beta_1$$

$$d_2 = \left(\frac{1}{2}m_2 + m_3\right) L_1 L_2 \cos\left(\beta_2 - \beta_1\right) \ddot{\beta}_1 - \left(\frac{1}{2}m_2 + m_3\right) L_1 L_2 \sin\left(\beta_2 - \beta_1\right) \dot{\beta}_1^2$$
$$+ \left(\frac{1}{2}m_2 + m_3\right) g L_2 \cos\beta_2.$$

$$\tag{3.34}$$

3.2.2 FRACTIONAL-ORDER PID CONTROLLER AND AUGMENTED SYSTEM STABILITY

In Equation (3.33), for the PID controller, the control variable V_k is expressed as:

$$V_k = K_{p,k} \left(\theta_k^d - \theta_{mk} \right) + K_{D,k} \left(\dot{\theta}_k^d - \dot{\theta}_{mk} \right) + K_{I,k} \int \left(\theta_k^d - \theta_{mk} \right) dt, \quad k = 1, 2, \quad (3.35)$$

where $K_{p,k}$, $K_{D,k}$, and $K_{I,k}$ are the PID proportional, derivative, and integral constants, respectively.

The fractional-order PID controller, denoted as $PI^\lambda D^\mu$, has five design parameters: proportional, derivative, integral constants, fractional order for the derivative order term, and fractional order for the integral order term, as shown in Equation (3.36):

$$V_k = K_{p,k} \left(\theta_k^d - \theta_{mk} \right) + K_{D,k} \, {}_0^C D_t^\mu \left(\dot{\theta}_k^d - \dot{\theta}_{mk} \right) + K_{I,k} \, {}_0^C D_t^{-\lambda} \left(\theta_k^d - \theta_{mk} \right). \quad (3.36)$$

For the trajectory tracking, θ_k^d is the desired angle, and θ_{mk} is the actual angle. In this case, the fractional derivative is Caputo's fractional derivative.

The closed-loop fractional-order PID control system of a robot arm is given in Figure 3.5.

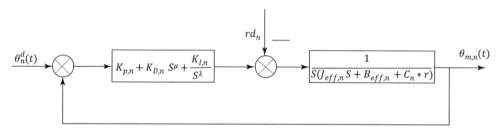

Figure 3.5: Closed loop diagram of fractional-order $PI^\lambda D^\mu$ controlled robot arm.

The corresponding closed loop transfer function is as follows, for simplicity, $\lambda, \mu \in (0, 1)$:

$$\theta_{m,n} = \frac{\left(K_p \theta_n^d - rd_n \right) s^\lambda + K_{I,n} \theta_n^d}{J_{eff,n} s^{\lambda+2} + \left(B_{eff,n} + C_n r \right) s^{\lambda+1} + K_{p,n} s^\lambda + K_{I,n}}. \quad (3.37)$$

Wan, Zhang, and French [2011] reported that the system is controllable and observable, and for a fractional-order system, the multilink robot system has significantly augmented stability, if all the eigenvalues of the dynamic system satisfy the following criteria, as shown in Figure 3.6.

3.2.3 NUMERICAL SIMULATION

The trajectory planning of the planar $x-y$ position, velocity, and acceleration of the multilink robot gripper is illustrated in Figure 3.7. The numerical simulations of the trajectory tracking using ordinary PID controller vs. fractional controller $PI^\lambda D^\mu$ are demonstrated in Figure 3.8, and the latter is proved to have superior performance.

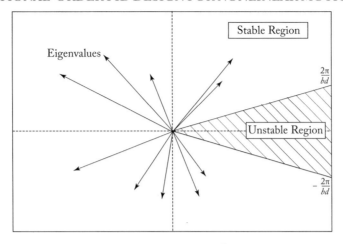

Figure 3.6: Stable region of the fractional controller $PI^\lambda D^\mu$ of the multilink robot.

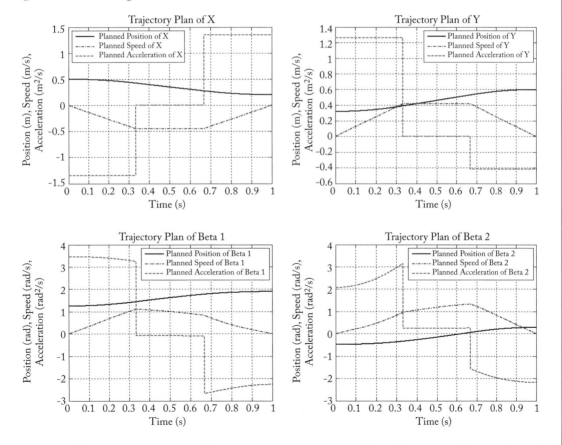

Figure 3.7: Trajectory planning of the multilink robot gripper (Wan, Zhang, and French [2011]).

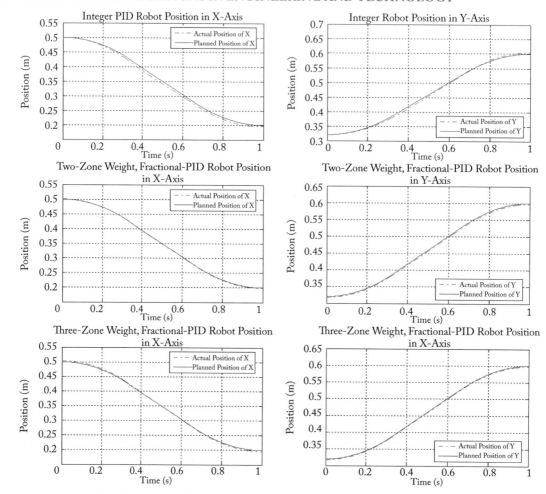

Figure 3.8: Trajectory planning of the multilink robot gripper (Wan, Zhang, and French [2011]).

3.3   FRACTIONAL-ORDER CONTROL FOR THE ROTARY INVERTED PENDULUM SYSTEM

The tracking control of the rotary arm and the stability control of the inverted pendulum can be achieved with a fractional-order PID controller. Preprocessed by input-output feedback linearization (IOFL), the rotary inverted pendulum (RIP) system is converted into a linearized pendulum-arm subsystem plus a nonlinear rotary-arm subsystem. The fractional-order PID (FOPID) controller is optimized using particle swarm optimization (PSO) algorithm to make the rotary arm track the reference signal. While the linearized subsystem is stabilized directly by a pole placement feedback controller.

3.3.1   MODELING OF THE ROTARY INVERTED PENDULUM SYSTEM

The RIP system is a typical electrical-mechanical system for controller design. The derivation of the system dynamics was presented in Yang and Zhang [2018]. The complete system dynamics can be excerpted as follows:

$$
\begin{aligned}
\left(J_r + m_1 l_{1c}^2 + C_2 \sin^2 \alpha + I_{2xx} \cos^2 \alpha + m_2 l_1^2\right) \ddot{\theta} + 2\left(C_2 - I_{2xx}\right) \dot{\theta} \dot{\alpha} \sin \alpha \cos \alpha \\
+ C_3 \ddot{\alpha} \cos \alpha - C_3 \dot{\alpha}^2 \sin \alpha = -b_\theta \dot{\theta} + \tau_m + \tau_{C\theta}
\end{aligned}
\tag{3.38}
$$

$$
C_1 \ddot{\alpha} + C_3 \ddot{\theta} \cos \alpha - \left(C_2 - I_{2xx}\right) \dot{\theta}^2 \sin \alpha \cos \alpha - C_4 \sin \alpha = -b_\alpha \dot{\alpha} + \tau_{C\alpha},
\tag{3.39}
$$

where $J_r, m_1, l_{1c}, I_{2xx}, m_2 l_1, b_\theta, \tau_{C\theta}, b_\alpha, \tau_{C\alpha}, C_1, C_2, C_3, C_4, C_5,$ and C_6 are coefficients determined from system identification. The single input to the system is motor torque τ_m, and the two outputs of the system are the rotary arm angle position θ and the pendulum arm angle position α. The system dynamics in Equations (3.38) and (3.39) can be expressed in a compact matrix form, as given by

$$
\begin{bmatrix}
\dot{\theta} \\
\ddot{\theta} \\
\dot{\alpha} \\
\ddot{\alpha}
\end{bmatrix}
=
\begin{bmatrix}
\dot{\theta} \\
\frac{1}{\det(M)}\left(C_1 \tau_m + a - b\right) \\
\dot{\alpha} \\
\frac{1}{\det(M)}\left(-C_3 \cos \alpha \tau_m + c - d\right)
\end{bmatrix}
\tag{3.40}
$$

$$
M = \begin{bmatrix}
C_2 \sin^2 \alpha + C_5 + I_{2xx} \cos^2 \alpha & C_3 \cos \alpha \\
C_3 \cos \alpha & C_1
\end{bmatrix},
\tag{3.41}
$$

where the new parameters in Equations (3.40) and (3.41) are given as

$$
a = C_1 \left(\left(-b_\theta - C_6 \sin 2\alpha \dot{\alpha}\right) \dot{\theta} + C_3 \sin \alpha \dot{\alpha}^2 + \tau_{C\theta}\right)
$$

$$
b = C_3 \cos \alpha \left(\frac{C_6}{2} \sin 2\alpha \dot{\theta}^2 - b_\alpha \dot{\alpha} + C_4 \sin \alpha + \tau_{C\alpha}\right)
$$

$$
c = \left(C_5 + I_{2xx} \cos^2 \alpha + C_2 \sin^2 \alpha\right) \cdot \left(C_4 \sin \alpha + \frac{C_6}{2} \sin 2\alpha \dot{\theta}^2 - b_\alpha \dot{\alpha} + \tau_{C\alpha}\right)
$$

$$
d = C_3 \cos \alpha \left(\left(-b_\theta - C_6 \sin 2\alpha \dot{\alpha}\right) \dot{\theta} + C_3 \sin \alpha \dot{\alpha}^2 + \tau_{C\theta}\right).
$$

3.3.2   FRACTIONAL CONTROL LAW DESIGN

Based on IOFL techniques, the system dynamics in Equations (3.40) and (3.41) can be divided into a linear subsystem and a nonlinear subsystem. That is, if an output transformation is applied as follows:

$$
\tau_m = \frac{\det(M)v - c + d}{-C_3 \cos \alpha}.
\tag{3.42}
$$

The subsystem associated with α is linearized, and the stabilization of the pendulum arm can resort to the pole placement method. The feedback control law to the linear subsystem is given by

$$v_1(t) = -K_3\alpha - K_4\dot{\alpha}. \tag{3.43}$$

However, when this control is plug into the nonlinear subsystem dynamics (θ), the goal of making the rotary arm track a reference signal can not be achieved. Thus, a second term can be added to Equation (3.43) to guarantee the simultaneous stabilization of the pendulum arm and the tracking control of the rotary arm. To promise the effectiveness and the robustness of the controller, the fractional-order PID controller is considered [Liao et al., 2015]. Therefore, the overall control law is given as

$$v(t) = v_1(t) + K_p e_1(t) + K_i D^{-\lambda} e_1(t) + K_d D^{\mu} e_1(t), \tag{3.44}$$

where $e_1(t)$ is the difference error between the rotary angle and the reference signal. K_p, K_i, K_d, λ, and μ are the fractional-order PID controller parameters. Even though this new control law also considers the nonlinear subsystem of θ, it should still be noted that the stability of the linear subsystem may lose due to the added fractional-order terms in the new control law. To overcome this drawback, the feedback control gains K_3 and K_4 should be determined in such a way that the poles of the closed-loop linear subsystem should be placed far from the imaginary axis on the open left half complex plane. The far-away placement of the poles for the linear subsystem will reduce the added terms' impacts on the pole-placement control of the linear subsystem. In this case, $K_3 = 800$ and $K_4 = 60$.

In addition, the optimization procedures to tune the fractional-order PID controller parameters are given in Yang, Zhang, and Voyles [2019]. The control goal for the nonlinear subsystem is to make the rotary angle track a reference signal, the cost function to be minimized is designed as follows:

$$J(K) = w_1 M_p + w_2 E_{ss} + w_3(t_s - t_r) + w_4 \int_0^T e_1^2(t)dt + w_5 \int_0^T t e_1^2(t)dt, \tag{3.45}$$

where performance indicators of the nonlinear subsystem, such as overshoot M_p, steady-state error E_{ss}, settling time t_s, rising time t_r, and the state error $e_1(t)$ of the linear subsystem are summed with weights to evaluate the complete cost. w_1, w_2, w_3, w_4, and w_5 are empirically selected weight coefficients, which are set as $w_1 = 10$, $w_2 = 1$, $w_3 = 0.5$, $w_4 = 1000$, and $w_5 = 200$ in Yang, Zhang, and Voyles [2019]. The last two terms of Equation (3.45) are the integral squared error and the integral timed squared error, which account for the cost of the tracking control for the nonlinear subsystem. This optimization problem is formally defined as

$$\tilde{K} = \arg\min_{K \in G} J(K). \tag{3.46}$$

The feasible set within which the optimization is functioning is specified by $G = \{K_p \in [5, 15], K_i \in [0, 5], K_d \in [5, 15], \in [0.5, 0.999], \in [0.5, 0.999]\}$. In order to solve this optimization problem, the PSO algorithm is applied. In PSO simulation, a set of tentative solutions are randomly selected from the feasible domain. After each iteration, the tentative solutions in the set will approach a little closer to their neighboring local optimizer. With many steps of iteration, the optimization simulation will converge to a global optimizer by choosing the best one from the set of local optimizers. The optimal parameters determined from the PSO simulation is determined as follows:

$$K_p = 10.017 \quad K_i = 0.104 \quad K_d = 12.277 \quad \lambda = 0.864 \quad \mu = 0.852.$$

As a result, the fractional-order operators in the $PI^\lambda D^\mu$ controller are approximated with the Oustaloup filter and given as follows:

$$s^{-0.864} \approx \frac{0.0015034s(s + 1111)(s + 877.7)(s + 128.8)}{(s + 596.1)(s + 167.2)(s + 24.55)(s + 3.603)} \cdot \frac{(s + 18.91)(s + 2.775)(s + 0.4074)(s + 0.0598)}{(s + 0.5289)(s + 0.07762)(s + 0.01139)(s - 0.0007776)} \tag{3.47}$$

$$s^{0.852} \approx \frac{2222s(s + 1111)(s + 169.2)(s + 24.83)}{(s + 7508)(s + 867.6)(s + 127.4)(s + 18.69)} \cdot \frac{(s + 3.645)(s + 0.535)(s + 0.07852)(s + 0.01153)}{(s + 2.744)(s + 0.4027)(s + 0.05911)(s + 0.0007668)} \tag{3.48}$$

3.3.3 NUMERICAL SIMULATION

The system responses are composed of the θ response, $\dot{\theta}$ response, α response, and $\dot{\alpha}$ response. The θ response and $\dot{\theta}$ response are the outputs of the linear subsystem, and the α and the $\dot{\alpha}$ are the outputs of the nonlinear subsystem. In order to evaluate the performance of the fractional-order control law in Equation (3.44), the simulation results from a similar but integer-order control law [Yang and Zhang, 2018] are compared with the simulation results from the fractional-order control law in Equation (3.44). The results are presented in Figures 3.9–3.12.

Figures 3.9 and 3.10 are the simulation responses with a step input signal. Figure 3.9 presents the simulation results for the nonlinear subsystem; the rotary arm angle should track a step input signal. Observed from the theta response, the system under the fractional-order control law successfully achieve the tracking goal set at the beginning, but the integer-order control law fails by forcing the theta angle to the zero position. Figure 3.10 is the simulation results of the linear subsystem, aiming at stabilizing the pendulum arm. It is revealed from the simulation results that both types of control laws can effectively stabilize the pendulum arm.

Figures 3.11 and 3.12 show the simulation responses with a sinusoidal input signal. It is displayed in Figure 3.11 that the fractional-order control law does ensure the rotary arm's tracking of a sinusoidal signal. However, the integer-order control law does not drive the rotary arm to track the reference signal. Even though the fractional-order control law and the integer-order control law have different control performance when they are applied for the tracking

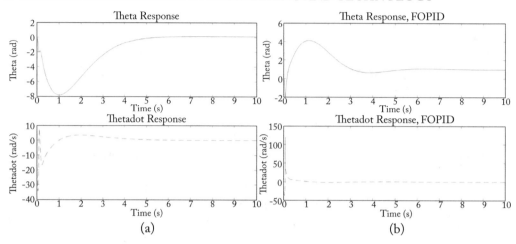

Figure 3.9: The system response of the rotary arm (nonlinear subsystem) with step input and (a) integer-order or (b) fractional-order control law.

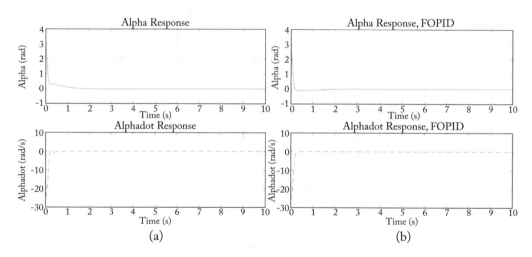

Figure 3.10: The system response of the pendulum arm (linear subsystem) with step input and (a) integer-order or (b) fractional-order control law.

control of the rotary arm, they do simultaneously guarantee the stabilization of the pendulum arm, which can be seen from the simulation results in Figure 3.12.

As per the simulation study of the rotary inverted pendulum system, the fractional-order control law not only achieves the stabilization of the pendulum arm, but also realizes the tracking control for the rotary arm subsystem. This example sums up the superiority of the fractional-order control schemes over the integer-order ones in the engineering and technology.

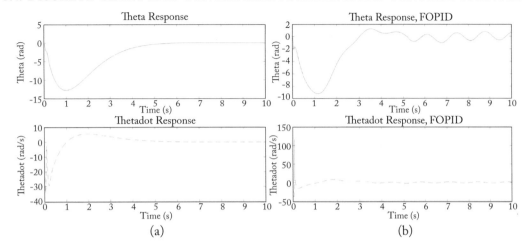

Figure 3.11: The system response of the rotary arm (nonlinear subsystem) with sinusoidal input and (a) integer-order or (b) fractional-order control law.

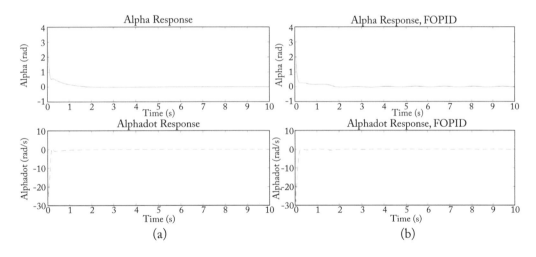

Figure 3.12: The system response of the pendulum arm (linear subsystem) with sinusoidal input and (a) integer-order or (b) fractional-order control law.

3.4   DISCRETE-TIME FOPID CONTROLLER FOR IDLE SPEED CONTROL OF AN ICE

FOPID controller has been applied to the tracking control of the RIP system. In this section, a discretized FOPID controller is developed to reduce the external disturbance's influence on the idle speed system of an internal combustion engine (ICE) and maintain the constancy of the idle speed. The linearized idle speed dynamics will be approximated by a first order plus dead time (FOPDT) model, and then, the FOPID controller parameters can be initialized by applying a Nichols–Ziegler type tuning rule to the FOPDT plant [Valerio and Costa, 2005]. This FOPID controller with the parameters tuned by Nichols–Ziegler type rule can stabilize the linearized idle speed system, but it may fails for the nonlinear idle speed dynamics. Thus, the controller parameters can be optimized within the vicinity of the initial parameter values to minimize a specified cost function. The genetic algorithm (GA) will be taken in the optimization. At the end of this section, the simulation results from the optimized discrete-time FOPID controller will be compared with the results from the optimized discrete-time FOPID controller [Yang et al., 2019].

3.4.1   MODELING OF THE IDLE SPEED SYSTEM OF AN ICE

The idle speed system model can be derived with four steps: the air flow dynamics in the throttle, the manifold filling dynamics, engine combustion dynamics, and the engine crankshaft dynamics. Its mathematical formulation are given in Equations (3.49)–(3.52) [Hrovat and Sun, 1997, Nicolao et al., 1999, Zhang, 2007].

(1) The air flow dynamics in the throttle:

$$\dot{m}_{th} = \frac{\tilde{A}_{th} P_0}{\sqrt{R T_0}} \sqrt{\gamma \left(\frac{2}{\gamma + 1} \right)^{\frac{\gamma + 1}{\gamma - 1}}}. \tag{3.49}$$

(2) The manifold filling dynamics:

$$\dot{P}_m = \frac{R T_m}{V_m} \left[\dot{m}_{th} - \dot{m}_m + \frac{P_m V_m}{R T_m^2} \dot{T}_m \right]. \tag{3.50}$$

(3) Engine combustion dynamics:

$$T_{ind} = \eta_{sp} \eta_{AF} \left[c_1 P_m + c_0 \right]. \tag{3.51}$$

(4) Engine crankshaft dynamics:

$$T_{eng} - T_{friction} - T_{load} = J \dot{\omega}, \tag{3.52}$$

where ω is the output idle speed and α is input throttle angle, which determines the equivalent throttle operating area $\tilde{A}_{th}$. The other variables and parameters in Equations (3.49)–(3.52) are also defined in detail.

Based on the nonlinear dynamics of the idle speed system, perturbing the variables around their nominal values will produce the linearized idle speed dynamics of an ICE, which is given as follows:

$$\frac{d\delta P_m}{dt} = A_1\delta P_m + A_2\delta N + A_3\delta\alpha \tag{3.53}$$

$$\frac{d\delta N}{dt} = B_1\delta N + B_2\delta\beta + B_3\delta P_m(t - T_d), \tag{3.54}$$

where A_1, A_2, A_3, B_1, B_2, and B_3 are constant coefficients, T_d is the time delay from the indicated torque to the engine torque.

3.4.2   FRACTIONAL-ORDER CONTROLLER DESIGN

Given the linearized system dynamics in Equations (3.53) and (3.54), Yang et al. [2019] presented an approximated FOPDT plant, which is given in Equation (3.55):

$$\frac{N(s)}{A(s)} \approx \frac{967.1}{1 + 0.0496s}e^{-0.075s}. \tag{3.55}$$

Applying the Nichols–Ziegler-type tuning rule, the FOPID controller parameters can be initialized to be $K_p = 0.0397$, $K_i = 0.3503$, $K_d = 0.0539$, $\lambda = 1.5548$, and $\mu = 0.9005$. Let the optimization feasible set be a small neighboring region around the initial controller parameter values, that is $G = \{0 < K_p < 0.1, 0 < K_i < 0.5, 0 < K_d < 0.1, 0 < \lambda < 2, 0 < \mu < 2\}$. Then, the optimal values of the controller parameters can be solved to minimize a cost function, which is defined as follows:

$$\mathcal{J} = w_1 \int_0^{30} e^2(t)dt + w_2 \int_{10+\epsilon}^{20-\epsilon} e^2(t)dt. \tag{3.56}$$

Solving this optimization with GA algorithm will generate a set of local optimizer in the feasible region. By fixing the derivative and the integral order to be 1, a set of local optimizer for the integer-order PID controller parameters can be solved; see Table 3.2.

Table 3.2: Optimized parameters for FOPID and PID.

Controller	K_p	K_i	K_d	λ	μ
PID	0.01003	0.06702	0.00107	1	1
FOPID	0.00957	0.06092	0.00124	0.97854	1.3001

Last, discretize the FOPID controller using the power series expansion and Euler's rule, the discrete-time FOPID controller and the discrete-time PID controller are produced [Merrikh-Bayat, Mirebrahimi, and Khalili 2014].

3.4.3   NUMERICAL SIMULATION

The simulation model are set up to compare the performance of the discrete-time FOPID controller and the performance of the discrete-time PID controller under the square-shape external disturbance. The simulation results are recorded in Figures 3.13 and 3.14. It is displayed in Figure 3.13 that the optimal FOPID controller can stabilize the idle speed with less settling time and overshoot than the optimal PID controller. Similarly, Figure 3.14 shows that less control efforts are needed if the optimal PID controller is replaced by the optimal FOPID controller.

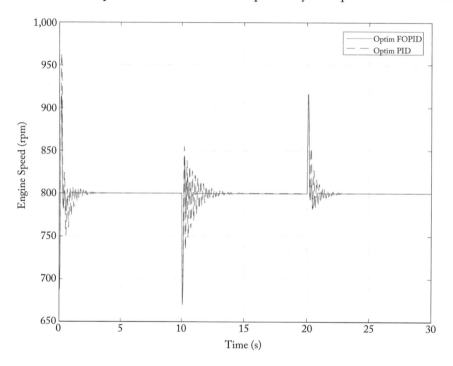

Figure 3.13: The idle speed response under a square-shape disturbance, controlled by an optimal FOPID controller or an optimal PID controller.

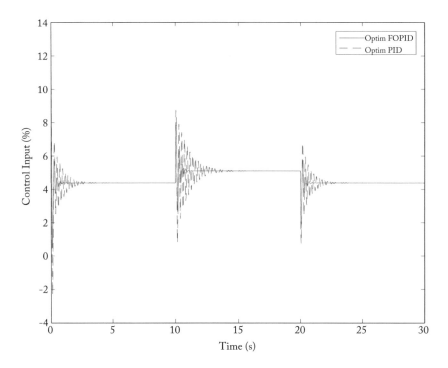

Figure 3.14: The control input (throttle angle α) under a square-shape disturbance, controlled by an optimal FOPID controller, or an optimal PID controller.

Bibliography

Balachandran, K., Govindaraj, V., Ortigueira, M. D., Rivero, M., and Trujillo, J. J. (2013). Observability and controllability of fractional linear dynamical systems. *IFAC Proceedings*, vol. 46(1), pp. 893–898. DOI: 10.3182/20130204-3-fr-4032.00081. 60

Benson, D. A., Meerschaert, M. M., and Revielle, J. (2013). Fractional calculus in hydrologic modeling: A numerical perspective. *Advances in Water Resources*, 51, pp. 479–497. DOI: 10.1016/j.advwatres.2012.04.005.

Caputo, M. (1969). *Elasticità e Dissipazione*. Bologna, Zanichelli. 69

Cheng, J. (2013). A note on the fractional Green's function. *Communications in Fractional Calculus*, pp. 16–24. 31

Chong, E. K. P. and Zak, S. H. (2013). *An Introduction to Optimization*, 4th ed., vol. 76, John Wiley & Sons. DOI: 10.1002/9781118033340. 64

Das, S. (2008). *Functional Fractional Calculus: For System Identification and Controls*, 1st ed., Springer, Mumbai, India. DOI: 10.1007/978-3-540-72703-3. 10

Dzieliński, A., Sarwas, G., and Sierociuk, D. (2011). Comparison and validation of integer and fractional order ultracapacitor models. *Advances in Difference Equations*. DOI: 10.1186/1687-1847-2011-11.

EqWorld. (2019). Mellin transforms: General formulas. *Retrieved from the World of Mathematical Equations*. http://eqworld.ipmnet.ru/en/auxiliary/inttrans/mellin.pdf 27

Folea, S., Keyser, R. D., Birs, I. R., Muresan, C. I., and Ionescu, C. (2017). Discrete-time implementation and experimental validation of a fractional order PD controller for vibration suppression in airplane wings. *Acta Polytechnica Hungarica*, 14(1), pp. 191–206. DOI: 10.12700/aph.14.1.2017.1.13.

Hrovat, D. and Sun, J. (1997). Models and control methodologies for IC engine idle speed control design. *Control Engineering Practice*, pp. 1093–1100. DOI: 10.1016/s0967-0661(97)00101-9. 86

Khalil, H. K. (2002). *Nonlinear Systems*. Prentice Hall, Upper Saddle River, NJ. 22

Kwakernaak, H. and Sivan, R. (1972). *Linear Optimal Control Systems*. Wiley Interscience, New York. DOI: 10.1115/1.3426828. 53

Lee, T. N. and Dianat, S. (1981). Stability of time-delay systems. *IEEE Transactions on Automatic Control*, pp. 951–953. DOI: 10.1109/tac.1981.1102755. 68

Liao, W., Liu, Z., Wen, S., Bi, S., and Wang, D. (2015). Fractional PID based stability control for a single link rotary inverted pendulum. *IEEE*, Beijing, China. DOI: 10.1109/icamechs.2015.7287174. 82

Lubich, C. (1986). Discretized fractional calculus. *SIAM Journal of Mathematical Analysis*, pp. 704-719. DOI: 10.1137/0517050. 36

Magin, R. (2006). *Fractional Calculus in Bioengineering*, 1st ed., Begell House. DOI: 10.1615/critrevbiomedeng.v32.10. 10

Merrikh-Bayat, F., Mirebrahimi, N., and Khalili, M. R. (2014). Discrete-time fractional-order PID controller: Definition, tuning, digital realization and some applications. *International Journal of Control Automation and Systems*, 13(1), pp. 81–90. DOI: 10.1007/s12555-013-0335-y. 87

Monje, C. A., Calderon, A. J., Vinagre, B. M., Chen, Y., and Feliu, V. (2004). On fractional PIλ controllers: Some tuning rules for robustness to plant uncertainties. *Nonlinear Dynamics*, 38(1–4), pp. 369–381. DOI: 10.1007/s11071-004-3767-3.

Monje, C. A., Chen, Y. Q., Vinagre, B. M., Xue, D., and Feliu, V. (2010). *Fractional-Order Systems and Controls: Fundamentals and Applications*. Springer-Verlag, London. DOI: 10.1007/978-1-84996-335-0. 6, 57

Nicolao, G. D., Rossi, C., Scattolini, R., and Suffritti, M. (1999). Identification and idle speed control of internal combustion engines. *Control Engineering Practice*, 7(9), pp. 1061–1069. DOI: 10.1016/s0967-0661(99)00051-9. 86

Oustaloup, A. (1995). *La Dérivation Non Entière: Théorie, Synthèse et Applications*. Hermes Science Publications, Paris. 43

Petás, Ivo. (2011). Fractional derivatives, fractional integrals, and fractional differential equations in Matlab. In A. H. Assi, *Engineering Education and Research Using MATLAB*, pp. 239–264, INTECH, Rijeka, Croatia. DOI: 10.5772/1532. 10

Podlubny, I. (1999). *Fractional Differential Equations*, 1st ed., vol. 198, Academic Press, San Diego. 4, 6, 8, 13, 19, 28, 32, 36

Raynaud, H. F. and Zergainoh, A. (2000). State-space representation for fractional order controllers. *Automatica*, 36(7), pp. 1017–1021. DOI: 10.1016/s0005-1098(00)00011-x. 56

Singhal, R., Padhee, S., and Kaur, G. (2012). Design of fractional order PID controller for speed control of DC motor. *International Journal of Scientific and Research Publications*, 2(6), pp. 1–8.

Slotine, J.-J. E. and Li, W. (1991). *Applied Nonlinear Control*. Prentice Hall, Englewood Cliffs, NJ. 22

Türker, T., Görgün, H., and Cansever, G. (2012). Lyapunov's direct method for stabilization of the Furuta pendulum. *Turkish Journal of Electrical Engineering and Computer Sciences*, 20(1), pp. 99–110.

Uchaikin, V. (2013). *Fractional Derivatives for Physicists and Engineers*, vol. I, Background and Theory. Higher Education Press, Beijing. DOI: 10.1007/978-3-642-33911-0. 4

Valerio, D. and Costa, J. S. (2005). Ziegler–Nichols type tuning rules for fractional PID controllers. *ASME International Design Engineering Technical Conferences and Computers and Information in Engineering Conference*, pp. 1–10, Long Beach, CA. DOI: 10.1115/detc2005-84344. 86

Wan, Y., Zhang, H. H., and French, M. (2011). Fractional order PID design for nonlinear motion control based on adept 550 robot. *International Journal of Modern Engineering*, 12(1), pp. 11–19. 78, 79, 80

Wan, Y., Zhang, H. H., and French, M. (2015). Adjustable fractional order adaptive control on single-delay regenerative machining chatter. *Journal of Fractional Calculus and Applications*, 6(1), pp. 185–207. 70, 73

Wang, F. S., Juang, W. S., and Chan, C. T. (1995). Optimal tuning of PID controllers for single and cascade control loops. *Chemical Engineering Communications*, pp. 15–34. DOI: 10.1080/00986449508936294. 53

Wheeden, R. L. and Zygmund, A. (2015). *Measure and Integral: An Introduction to Real Analysis*, 2nd ed., Chapman and Hall/CRC, Boca Raton, FL. DOI: 10.1201/b18361. 28

Xue, D. (2017). *Fractional-Order Control Systems: Fundamentals and Numerical Implementations*, 1st ed., De Gruyter, Berlin, Boston. DOI: 10.1515/9783110497977. 6, 22, 43, 54

Xue, D. and Bai, L. (2016). Numerical algorithms for Caputo fractional-order differential equations. *International Journal of Control*, pp. 1–19. DOI: 10.1080/00207179.2016.1158419. 39

Yang, Y. and Zhang, H. H. (2018). Stability study of LQR and pole-placement genetic algorithm synthesized input-output feedback linearization controllers for a rotary inverted pendulum system. *International Journal of Engineering Innovations and Research*, 7(1), pp. 62–68. 81, 83

Yang, Y., Zhang, H. H., and Voyles, R. M. (2019). Rotary inverted pendulum system tracking and stability control based on input-output feedback linearization and PSO-optimized fractional order PID controller. *Proc. of the International Conference on Automatic Control,*

Mechatronics and Industrial Engineering (ACMIE), pp. 79–84, CRC Press, Suzhou, China, October 29–31. https://www.crcpress.com/Automatic-Control-Mechatronics-and-Industrial-Engineering-Proceedings/He-Qing/p/book/9781138604278 82, 86

Yang, Y., Zhang, H. H., Yu, W., and Tan, L. (2019). Optimal design of discrete-time fractional-order PID controller for idle speed control of an IC engine. Manuscript submitted for publication. 87

Zhang, H. H. (1996). Chatter modeling, analysis and control for CNC machining systems. Ph.D. dissertation, University of Michigan-Ann Arbor. 69

Zhang, H. H. (2007). Modeling and control of the idle speed dynamics of an IC engine. *Powertrain Control Project*, Ohio State University, Columbus, OH. 86

Zhang, H. H. (2010). PID controller design for a nonlinear motion control based on modeling the dynamics of adept 550 robot. *International Journal of Industrial Engineering and Management*, 1(1), pp. 19–27. 74

Authors' Biographies

YI YANG

Yi Yang is currently a Ph.D. student in the School of Engineering Technology at Purdue University. He received a B.S. in Mechanical Engineering from Huazhong University of Science and Technology in 2015, and an M.S. from The University of Michigan–Ann Arbor in 2017. Yi's research interest lies in the nonlinear system control and mechatronics design. He is also interested in developing computer programs to solve optimization problems in industry and academia.

HAIYAN HENRY ZHANG

Haiyan Henry Zhang is a Full Professor of Engineering Technology and the founding director of Center for Technology Development at Purdue University. He received his Ph.D. from The University of Michigan–Ann Arbor in 1996. With the multidisciplinary engineering background and industrial experience, his research foci are analytical mechatronic design, advanced manufacturing control, and design of traditional or hybrid electric vehicle powertrains.

Printed in the United States
by Baker & Taylor Publisher Services